Future Climate of the Mediterranean and Europe

BY

ALEXANDRE TUEL
University of Bern
and
ELFATIH A. B. ELTAHIR
Massachusetts Institute of Technology

Table of contents

Edited by Stephanie M. McPherson

2022

Front cover image depicts projected changes in precipitation and sea-level pressure due to changing climate

Chapter 1: Introduction

Alexandre Tuel

Climate change will have major impacts on the Earth's hydrologic cycle. As greenhouse gas concentrations continue to rise, precipitation—which includes rain, snow, sleet and hail—will increase on average by 2% per each degree centigrade of additional warming. Yet, this average hides large differences across space. Higher precipitation is indeed expected over most of the world's continents, but in a few regions, climate models project robust declines. Among these, the Mediterranean stands out as a result of the magnitude and significance of its expected precipitation decline. Locally, up to 40% of annual precipitation could vanish. The decline is confined to the southern half of the basin during winter, but extends northwards during spring and summer. This is especially worrying as the region is already water-stressed. Droughts and wildfires are increasingly frequent all around the Mediterranean coastline. Any future decline in precipitation will set strong limits on water resources that will constrain the ability of the region to develop and grow food, affecting millions of people and threatening the stability of this tense and complex area.

This book brings together research that explains why the Mediterranean region is a climate change hot spot and how climate projections translate at the regional scale. We begin by discussing the physical drivers behind the projected winter drying in Chapters 2 and 3, and then discuss the implications for the climate of several regions in the southern Mediterranean in Chapters 4, 5 and 6. In the last chapter, we explain trends in spring and summer precipitation and further discuss their implications.

In Chapter 2, we take a new look at the physical drivers of the projected wintertime decline in precipitation. Regional circulation changes, dominated by the development of a strong anomalous ridge, are thought to drive the winter precipitation decline, but their origins and potential contributions to regional hydroclimate change remain elusive. Here, we show how wintertime Mediterranean circulation trends can be seen as the combined response to two independent forcings: robust changes in large-scale, upper-tropospheric flow and the reduction in the regional land–sea temperature gradient that is characteristic of this region. In addition, we discuss how the circulation change can account for the magnitude and spatial structure of the drying.

Chapter 3 further develops the results of the previous chapter and provides new arguments for why and how the Mediterranean is projected to experience large atmospheric circulation changes during winter. First, we find that zonal asymmetries in the vertical structure of stationary waves are important to explain the enhanced circulation response in the region and that these asymmetries are related through the external mode to the vertical structure of the mean zonal wind.

Second, in winter, the Mediterranean is located just to the north of the Hadley cell edge and consequently is relatively free of large-scale descent; together with low near-surface static stability above the sea, this condition allows the weaker warming trend above the sea to propagate to the low troposphere and trigger a major circulation response. During summer, however, remotely forced descent and strong static stability prevent the cooling anomaly from expanding upward. Most of the intermodel scatter in the projected low-level circulation response in winter is related to the spread in upper-tropospheric dynamical trends. An important finding in this chapter is that because climate models exhibit too much vertical coherence over the Mediterranean, they likely overestimate the sensitivity of Mediterranean near-surface circulation to large-scale dynamical changes.

Chapters 4 through 6 then focus on the Southwestern and Eastern Mediterranean, two regions which will experience the largest declines in winter precipitation. In Chapter 4, we evaluate the performance of the MIT Regional Climate Model (MRCM) in simulating seasonal and inter-annual regional climatic features over Northwestern Africa and the Iberian Peninsula (NWAIP), and develop and interpret regional climate change projections over the same region under the RCP4.5 and 8.5 scenarios, at a resolution of 12 km, driven by three carefully selected GCMs. MRCM succeeds in correctly capturing the spatio-temporal patterns of precipitation when forced by ERA-Interim reanalysis data. In addition to substantial year-round warming, the MRCM ensemble projects a significant decline in precipitation during winter and spring under continued anthropogenic forcing. Even with significant climate change mitigation, the drying is relatively large, with an average of − 34% in precipitation over Morocco's three main watersheds during spring, and − 22% during winter.

Projections for the Iberian Peninsula are less severe, though drying trends are still robust. The consistency of the precipitation projections is discussed in the light of physical mechanisms acting at the regional scale. Being located downwind of a large desert, the region is particularly vulnerable to the changes in circulation projected by global models. The latter will advect dry air from the Sahara, which will suppress precipitation. The local topography also contributes to these trends. The results presented in this chapter elucidate the physical processes responsible for winter and spring drying in the Southwestern Mediterranean.

The High Atlas, culminating at more than 4000 m, is the water tower of Morocco. While plains receive less than 400 mm of precipitation in an average year, the mountains can get twice as much, often in the form of snow between November and March. Snowmelt thus accounts for a large fraction of the river discharge in the region, particularly during spring. In parallel, future climate change projections point towards a significant decline in precipitation and enhanced warming of temperature for the area. In Chapter 5, we build on previous research results on snow and climate modelling in the High Atlas to make detailed projections of snowpack and river flow response to climate change in this region. We develop end-of-century snowpack projections using a distributed energy balance snow model based on SNOW-17 and high-resolution climate simulations over Morocco with the MIT Regional Climate Model (MRCM) under a mitigation (RCP4.5) and a business-as-usual (RCP8.5) scenario. Snowpack water content is projected to decline by up to 60 % under RCP4.5 and 80 % under RCP8.5 as a consequence of strong warming and drying in the region. We also implement a panel regression framework to relate runoff ratios to regional meteorological conditions in seven small sub-catchments in the High Atlas. Relative humidity and the fraction of solid-to-total precipitation are found to explain about 30% of the inter-annual variability in runoff ratios. Due to projected future atmospheric drying and the associated decline in snow-to-precipitation ratio, a 5 %–30 % decrease in runoff ratios and 10 %–60 % decrease in precipitation are expected to lead to severe (20 %–70 %) declines in river discharge.

The recent drought in the Eastern Mediterranean may have contributed to the deterioration in economic and social conditions that fueled recent conflicts in this unstable region. Chapter 6 details how future climate change may exacerbate dry conditions in this region. In this chapter, we evaluate the performance of the MIT Regional Climate Model (MRCM) in simulating the Eastern Mediterranean at a high resolution of 7 km (MRCM/7km), and develop climate change projections for this region under several greenhouse gas emission scenarios. The high resolution is reached by applying a double-nesting methodology relying on 25km MRCM simulations forced by three carefully selected global climate models (GCMs) as boundary conditions. Compared to GCMs and the coarser 25km simulation, MRCM/7km can simulate more accurately the regional-scale distribution of precipitation and temperature over this mountainous region compared to historical observations, particularly the sharp east-west gradient of rainfall characteristic of this region. Still, significant discrepancies persist compared to local station observations, which limit the validity of our projections at the local-scale of sub-regions and individual stations. Based on projected precipitation changes under both RCP4.5 and RCP8.5 scenarios, MRCM/7km indicates more intense future droughts in this region with a relatively uniform spatial pattern, which is significantly different from the westward increasing pattern projected by coarser-resolution models. The projected changes in precipitation are explained by future circulation changes, a shift towards more northerly winds, bringing cold and dry air from central Europe and the Anatolian Plateau. Forced descent downstream of orography results in severe reduction in precipitation and accounts for the robustness of the projections.

Finally, Chapter 7 moves away from winter to analyse the projected summertime decline in precipitation in the Northern Mediterranean and Europe. The geography of Europe as a continental landmass, located between the arid Sahara and the cold high latitudes (both are dry in terms of absolute humidity), dictates the reliance during summer of southern Europe (south of 458N) on stored water from winter and spring, and of northwestern Europe on a small concentrated low-level moisture jet from the North Atlantic. In this chapter, based on the analysis of observations and output from models from phase 5 of the Coupled Model Intercomparison Project, we expand this theory further, documenting how the winter precipitation decline expands into southern Europe during spring, dictated by similar dynamical mechanisms, depleting soil moisture and setting the stage for drier summers via soil moisture precipitation feedbacks. Over northwestern Europe, an anomalous anticyclonic circulation west of the British Isles displaces the low-level moisture jet northward, limiting moisture supply, and reducing low-level relative humidity and rainfall. Finally, we discuss how this comprehensive perspective of European summer climate change can help us better understand the variations across model projections, and pave the way for their reduction.

Chapter 2: Why Is the Mediterranean a Climate Change Hot Spot?

Alexandre Tuel
Elfatih A. B. Eltahir

Higher precipitation is expected over most of the world's continents under climate change, except for a few specific regions where models project robust declines. Among these, the Mediterranean stands out as a result of the magnitude and significance of its winter precipitation decline. Locally, up to 40% of winter precipitation could be lost, setting strong limits on water resources that will constrain the ability of the region to develop and grow food, affecting millions of already water-stressed people and threatening the stability of this tense and complex area. To this day, however, a theory explaining the special nature of this region as a climate change hot spot is still lacking. Regional circulation changes, dominated by the development of a strong anomalous ridge, are thought to drive the winter precipitation decline, but their origins and potential contributions to regional hydroclimate change remain elusive. Here, we show how wintertime Mediterranean circulation trends can be seen as the combined response to two independent forcings: robust changes in large- scale, upper-tropospheric flow and the reduction in the regional land–sea temperature gradient that is characteristic of this region. In addition, we discuss how the circulation change can account for the magnitude and spatial structure of the drying. Our findings pave the way for better understanding and improved modeling of the future Mediterranean hydroclimate.

Located at the border between the arid subtropics and the temperate midlatitudes, the Mediterranean Basin is characterized by low annual precipitation totals and high interannual variability, which impose a state of semipermanent water stress across much of North Africa and the Middle East. Summers are warm and dry, dominated to the east by the influence of subtropical remote forcing triggered by the Indian monsoon, which causes intense subsidence across the region (Rodwell and Hoskins 1996), and to the west by the subtropical high. In winter, however, the Mediterranean Sea region is largely outside the influence of such tropical teleconnections, and storms and rain are brought by midlatitude westerlies. Consequently, winter precipitation is key to the region's agriculture and economy, with its future of paramount importance for the basin's countries. Regional- and local-scale processes, such as land–sea circulations, also play a significant part in shaping Mediterranean climate variability and climate change (Bolle 2003).

The Mediterranean has long stood out in successive generations of global climate models (GCMs) as being particularly sensitive to rising concentrations of green- house gases. Models overwhelmingly project, across all scenarios, a large reduction in precipitation, more than in other land regions in relative terms (Figure 2.1b) (Giorgi and Lionello 2008; Planton et al. 2012). A large part of that decline occurs during winter, south of 40°N (Figure 2.1d), with enhanced drying over northwestern Africa (from -30% to -40% in December–February (DJF) precipitation) and the eastern Mediterranean (from -20% to -25%). In summer,

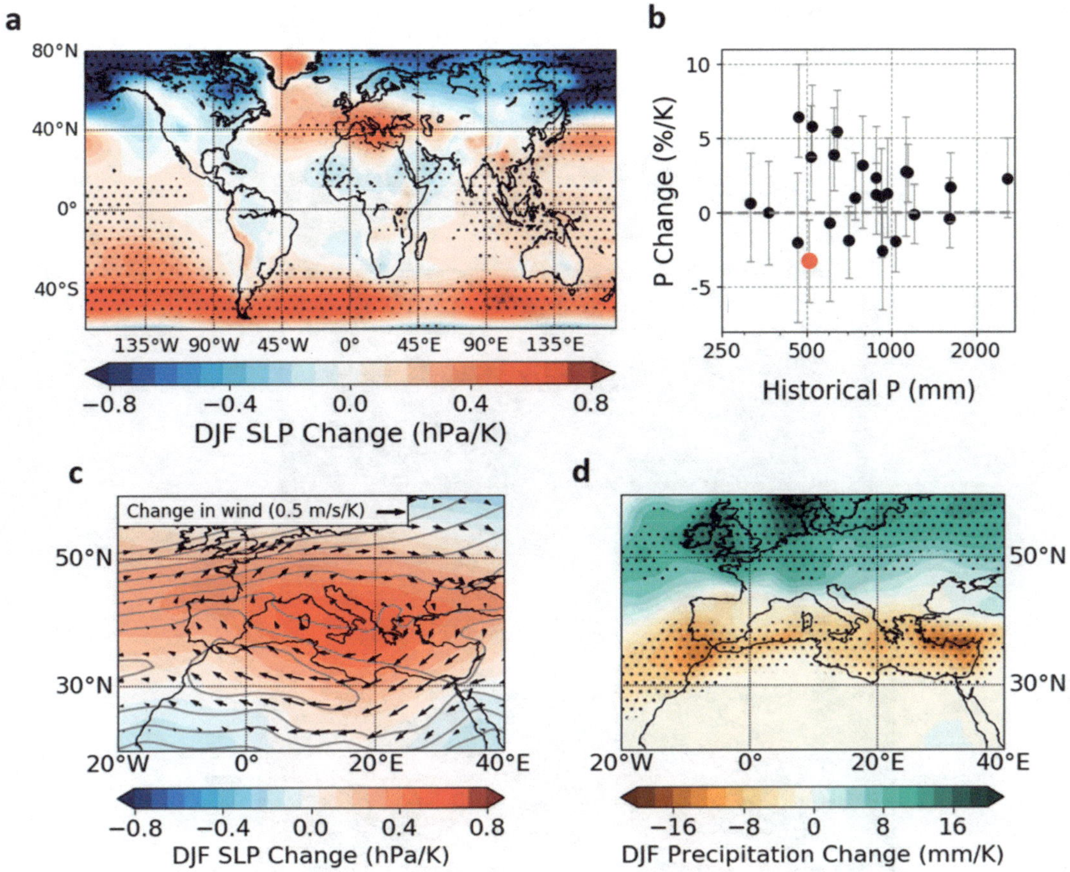

Figure 2.1: CMIP5 winter (DJF) multimodel mean projected change in (a) worldwide SLP, (c) Mediterranean SLP and 850-hPa winds, and (d) Mediterranean precipitation under RCP8.5 (2071–2100 minus 1976–2005). Historical DJF SLP contours are also shown in (c) (from 1010 to 1024 hPa every 2 hPa; the closed contour over North Africa is 1024 hPa). Dots in (a) and (d) indicate that more than 80% of models agree on the sign of the change. Also shown is (b) median (black dots) and 90% intermodel spread (gray whiskers) of relative change in annual precipitation in CMIP5 models under RCP8.5 (2071–2100 minus 1976–2005), against annual mean historical (1976–2005) precipitation, for 25 land regions covering the whole globe (Giorgi and Bi 2005). The Mediterranean is highlighted with a red dot. In (a)–(d), projections have been renormalized by each model's global projected temperature change.

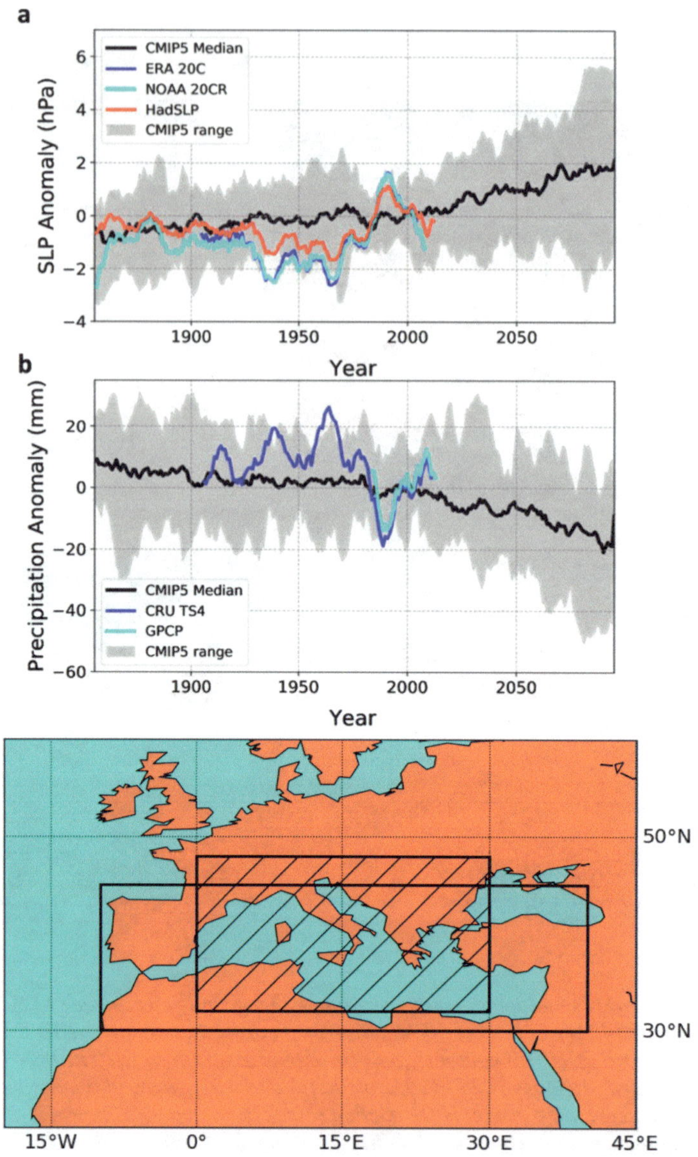

Figure 2.2: (Top) Five-year smoothed median (black line) and range (gray shading) of CMIP5 model simulations of DJF Mediterranean (a) SLP (0°–30°E, 32°–48°N) and (b) precipitation (20°W–40°E, 30°–45°N) anomalies (map on bottom), alongside observations/reanalysis data from HadSLP2, NOAA 20CR, and ERA-20C [in (a)] and GPCP and CRU TS4 v2 [in (b)]. (Below) Domains used to define Mediterranean-mean precipitation (20°W-40°E/30-45°N, un- hatched) and SLP (0-30°E/32-48°N, hatched).

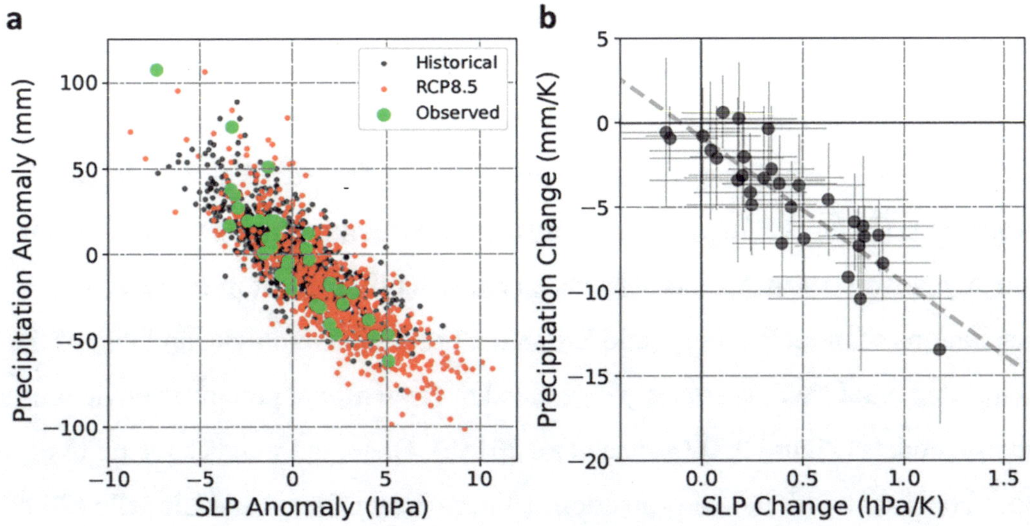

Figure 2.3: (a) DJF Mediterranean (20°W-40°E/30-45°N) precipitation against SLP (0-30°E/32-48°N) anomalies in historical (1976-2005) and RCP8.5 (2071-2100) CMIP5 simulations, and observations (HadSLP2/GPCP, 1979-2017). (b) Change in DJF Mediterranean precipitation and SLP across CMIP5 models, with 95% confidence intervals and least-squares best fit. Changes are re-scaled by each model's projected global temperature change.

significant warming and drying is also projected for the northern Mediterranean (Brogli et al. 2019). While the large interannual variability in wintertime Mediterranean climate makes the significance of past trends hard to establish (Kelley et al. 2012), observations and reanalysis products are in general consistent with historical simulations and projections for the upcoming century (Figure 2.2b). The Mediterranean has experienced substantial drying over the last century, part of which cannot be explained by simple internal variability (Hoerling et al. 2012; Kelley et al. 2012). An unprecedented drought in the eastern Mediterranean exacerbated the already tense situation in Syria that led to the outbreak of a civil war (Kelley et al. 2015). Further drying will inevitably exacerbate social and geopolitical tensions in this severely water-stressed region.

Changes in the regional low-level circulation have long been suspected to play a dominant role in the winter Mediterranean drying (e.g., Seager et al. 2014; Zappa and Shepherd 2017; Tuel and Eltahir 2018; Brogli et al. 2019), in particular through the development of an anomalous surface anticyclone over the Mediterranean Basin (Seager et al. 2019), a striking feature of global climate projections (Figures 2.1a,c). This anomalous Mediterranean ridge extends from roughly November to April and has been consistently present in successive generations of models (Giorgi and Lionello 2008). Its magnitude, and that of the associated wind field, is strongly correlated to the regional precipitation decline across models (Figure 2.3) (Zappa et al. 2015b). Observed trends have up to now been consistent with model projections (Figure 2.2a.) Giorgi and Lionello (2008) suggested that the anomalous high may drive the projected drying by increasing atmospheric stability in the region and suppressing Mediterranean cyclones, as discussed in later studies (e.g., Rojas et al. 2013; Zappa et al. 2015a).

Subsequently, based on a three- dimensional analysis of the moisture budget, Seager et al. (2014) concluded that the Mediterranean precipitation decline was due to increased moisture divergence by the time-mean flow, due to anomalous anticyclonic circulation in the region. Zappa et al. (2015b) additionally showed that the intermodel spread in precipitation change could be well captured by a simple wind index reflective of the regional circulation, supporting the idea that future Mediterranean hydro- climate trends are primarily driven by changes in the regional atmospheric circulation (Figure 2.3).

Why the anomalous high pressure develops in the first place, however, and how it connects to the robust pattern of the precipitation response remains unclear. Changes in the large-scale circulation, notably the expansion of the

Hadley circulation (Lu et al. 2007) and the corresponding poleward shift of the North Atlantic storm track (Yin 2005; Scheff and Frierson 2012; Woollings et al. 2012), the weakening of the Mediterranean storm track, and changes in regional flow regimes (Zappa et al. 2015a; Rojas et al. 2013), have previously been suggested as driving mechanisms behind Mediterranean circulation trends. More recently, the role of shifts in Northern Hemisphere stationary waves was shown to modulate precipitation projections over California (Simpson et al. 2016) and suspected to impact the Mediterranean as well (Seager et al. 2019). Tuel and Eltahir (2018) also first suggested that the regional warming contrast between land and sea could play a role. Still, the specific contributions of those mechanisms to projected Mediterranean climate trends and their spread has not been clearly quantified. Additionally, the role of Hadley cell and storm-track shifts has been challenged as they were found to occur, in abrupt- CO_2 experiments, on a much faster time scale than the Mediterranean drying (He and Soden 2017). The zonal symmetry in Hadley cell shifts is also at odds with the amplified Mediterranean sea level pressure (SLP) signal (Figure 2.1a), so that it is not obvious why the Mediterranean would be particularly sensitive in climate projections.

Therefore, while much attention has been given to this region, a comprehensive theory for the wintertime Mediterranean precipitation decline is still lacking. In this study, we seek to understand what drives projected wintertime low-level circulation and precipitation trends in the region, with winter defined as the DJF period. Based on results of simulations from phase 5 of the Coupled Model Intercomparison Project (CMIP5) and idealized simulations with the Massachusetts Institute of Technology (MIT) Regional Climate Model, we show that global-scale circulation changes originating near the tropopause and the independent response to changes in the sea–land surface temperature gradient over the Mediterranean Basin greatly contribute to future trends in the Mediterranean hydroclimate. In addition, we dis- cuss the physical connections between circulation and precipitation projections in the western and eastern Mediterranean. Our primary focus is the CMIP5 multimodel mean, but intermodel spread is also discussed in the light of the two proposed mechanisms.

Data

We analyze recent surface temperature, pressure, and precipitation trends in the Mediterranean region using various datasets. Observed surface temperatures are taken from the NOAA Merged Land–Ocean Surface Temperature Analysis (MLOST; Vose et al. 2012) (land and ocean; 1850–2018), and the CRU TS4.02 dataset (Harris et al. 2014) (land only; 1901–2018). Monthly precipitation over land from CRU TS4.02 is also used, and, for ocean coverage, we consider data from the Global Precipitation Climatology Project, version 2.3 (Adler et al. 2003), as well (1979–2018). Sea surface temperature (SST) data are taken from HadISST (Rayner et al. 2003) and ERSST v5 (Huang et al. 2017). SLP from the HadSLP2 dataset (1850–2018) is used (Allan and Ansell 2006); for purposes of comparison, we also look at SLP in the NCEP–NCAR (Kalnay et al. 1996), NOAA 20CR (Compo et al. 2011), and ERA twentieth-century (20c; Poli et al. 2016) reanalyses.

Future climate trends are analyzed using 30 GCM simulations from CMIP5 (Taylor et al. 2012), under the historical and representative concentration pathway 8.5 (RCP8.5) scenarios. We used all models that provided SLP, precipitation, surface temperature, and specific humidity, temperature, winds, and pressure velocity on pressure levels at monthly resolution for our reference (1976–2005) and future (2071–2100) periods. A detailed list of the 30 selected models can be found in Table 2.1 at the end of the chapter. For each model, only the r1i1p1 ensemble member is used. All model output is regridded to a common 1° × 1° grid. Unless specified, all changes under RCP8.5 are defined as the 2071–2100 minus 1976–2005 average, and all anomalies (of SLP, precipitation, etc.) are defined with respect to the 1976–2005 reference period.

Quantifying the SLP response to upper-level circulation change

CMIP5 GCMs robustly agree on the pattern of Northern Hemisphere upper-tropospheric circulation change under continued anthropogenic forcing: a strengthening of the midlatitude jet (Barnes and Polvani 2013) and associated shift in the pattern of quasi-stationary waves (Brandefelt and Krnich 2008) (Figure 2.4). A direct consequence of these changes is the development of an upper-tropospheric, anomalous anticyclonic circulation over the Mediterranean. Because of the generally equivalent barotropic character of winter stationary waves (Held et al. 2002), it is expected that such a change in upper-level flow would translate into anticyclonic circulation and higher pressure at low levels.

Methods

To quantify the impact of trends in upper-tropospheric flow on Mediterranean low-level circulation, we apply to each selected GCM an analog-based "dynamical adjustment" model. We give here a brief overview of the method presented in Deser et al. (2016) and refer to their paper for further mathematical details. We consider two physical fields, a predictor field $\mathbf{X} \in \mathrm{R}^{p}$, and a predictand field $\mathbf{Y} \in \mathrm{R}^{q}$, linked by some physical relationship (variability in **X** influences variability in **Y**); p and q refer to the space dimension. We assume that

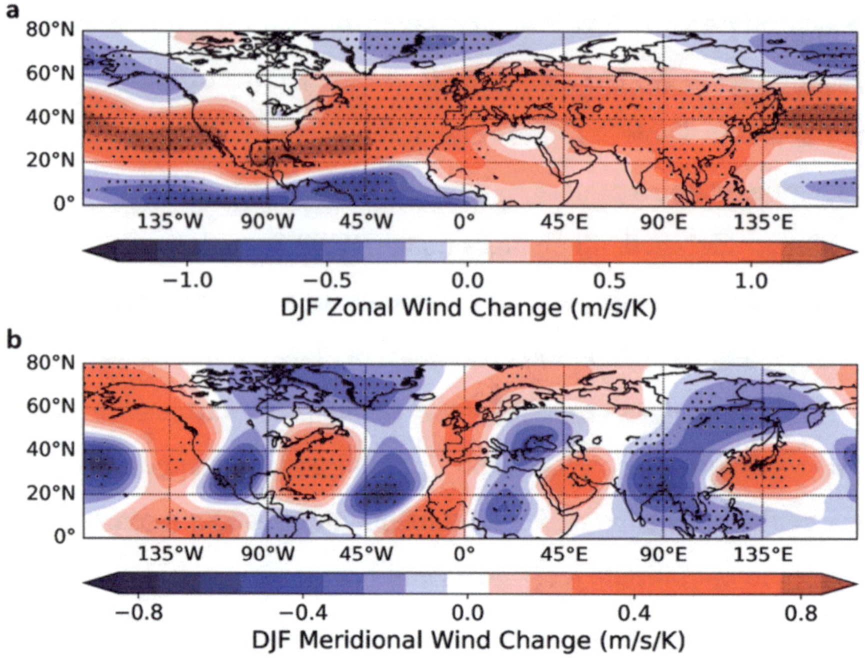

Figure 2.4: CMIP5 winter (DJF) multimodel mean projected change in Northern Hemisphere200-hPa (a) zonal and (b) meridional winds. Dots indicate that more than 80% of models agree on the sign of the change. For each model, projections have been renormalized by that model's global projected temperature change.

a "training" series of concurrent values of **X** and **Y**, of length n, is available, which we note $(\mathbf{X}_i)_{0\leq i\leq n}$ and $(\mathbf{Y}_i)_{0\leq i\leq n}$. The goal of dynamical adjustment is to estimate the amplitude of the field $\mathbf{Y}_t$ associated with the predictor field $\mathbf{X}_t$, observed at a time $t > n$. To that end, we select the $N = 50$ closest analogs of field $\mathbf{X}_t$ among the $(\mathbf{X}_i)_{0\leq i\leq n}$, with Euclidean distance used as the metric. Among these N fields, we randomly select a subsample of $M = 30$ fields, which are assembled in a $p \times M$ matrix $\mathbf{X}_c$. The corresponding **Y** values are similarly put in matrix $\mathbf{Y}_c$ Then, optimal linear com- bination weights b are estimated such that

$$\mathbf{X}_t \approx \beta \mathbf{X}_c \qquad (2.1)$$

$\beta \mathbf{X}_c$ represents a "constructed analog" of $\mathbf{X}_t$. The same weights β are also applied to $\mathbf{Y}_c$ to estimate the component of $\mathbf{Y}_t$ induced by predictor $\mathbf{X}_t$: $\widehat{\mathbf{Y}}_t = \beta \mathbf{Y}_c$. The procedure is then repeated 1000 times, each time with a new random subsample of 30 analogs. The purpose of this subsampling is to increase robustness of the results and better quantify the variability in the reconstructed estimates (Deser et al. 2016). The results are not particularly sensitive to the specific choice of N and M over the range $30 \leq N \leq 60$ and $M \approx 2N/3$.

The method is applied to each GCM separately, using for **X** the DJF meridional wind field anomalies at 200 hPa in the 20°–80°N latitude range, and for **Y** the Northern Hemisphere DJF SLP anomaly field. Seasonal anomalies are computed by subtracting from each field its 1976–2005 mean. The historical (1850–2005) simulations are used as "training" series (thus $n = 155$ winters), and we calculate constructed analogs for all winters between 2070 and 2100. Our choice of predictor is motivated by the results of Simpson et al. (2016), which implied that projected upper-level meridional wind anomalies were reflective of the shifts in mean stationary wave structure. Results are not significantly different when using the zonal wind component, or even the streamfunction, since upper-tropospheric flow is essentially nondivergent at seasonal time scales.

An important assumption of this approach is that, at first order at least, the surface circulation response to upper-level wind pattern change is linear. To test that hypothesis, we also reconstruct annual SLP anomalies in the historical runs: for each of the 155 winters in 1850– 2005, the other 154 winters are used to look for analogs. This helps determine whether interannual variability of Mediterranean SLP is correctly reconstructed in both historical and future simulations.

Results

Most of the year-to-year variability in DJF Mediterranean SLP is well reconstructed based on upper-tropospheric flow anomalies, in both historical and RCP8.5 runs (Figure 2.5c). The model-mean squared correlation coefficient is about 0.75 for each scenario (model range is 0.55–0.85), and the average root-mean-square error is 1.2 hPa (model range 0.95–1.5 hPa). SLP anomalies are correctly approximated over the whole range of interannual variability; only very low SLP values seem to be slightly overestimated. Despite the mean shift toward positive SLP anomalies in future projections, the range of values of RCP8.5 anomalies is not substantially different from historical ones. This increases our confidence that the method will correctly capture shifts in the distribution of SLP anomalies connected to projected changes in upper-level wind fields. Looking ahead, we find that future upper-tropospheric wind patterns are consistent with an amplified SLP response over the Mediterranean and also east of Japan (Figures 2.5a,b). They also account for 80% of intermodel spread in Mediterranean projections (Figure 2.5d). The decrease in SLP over the North Pacific is also well reproduced, but not so much over North America, the North Atlantic, and Siberia, where other factors, notably linked to thermodynamic forcing (e.g., Arctic amplification and the North Atlantic warming hole (NAWH)) likely play a role (Gervais et al. 2019). However, only 40% of the mean GCM response is accounted for (0.7 hPa as compared with 1.7 hPa in the multimodel mean). In some models, this dynamically induced SLP change is even of the opposite sign of the total simulated response by that same model (Figure 2.5d). Similarly, during the "extended" winter season, outside DJF, future upper-tropospheric circulation anomalies are still connected to an increase in SLP over the Mediterranean but

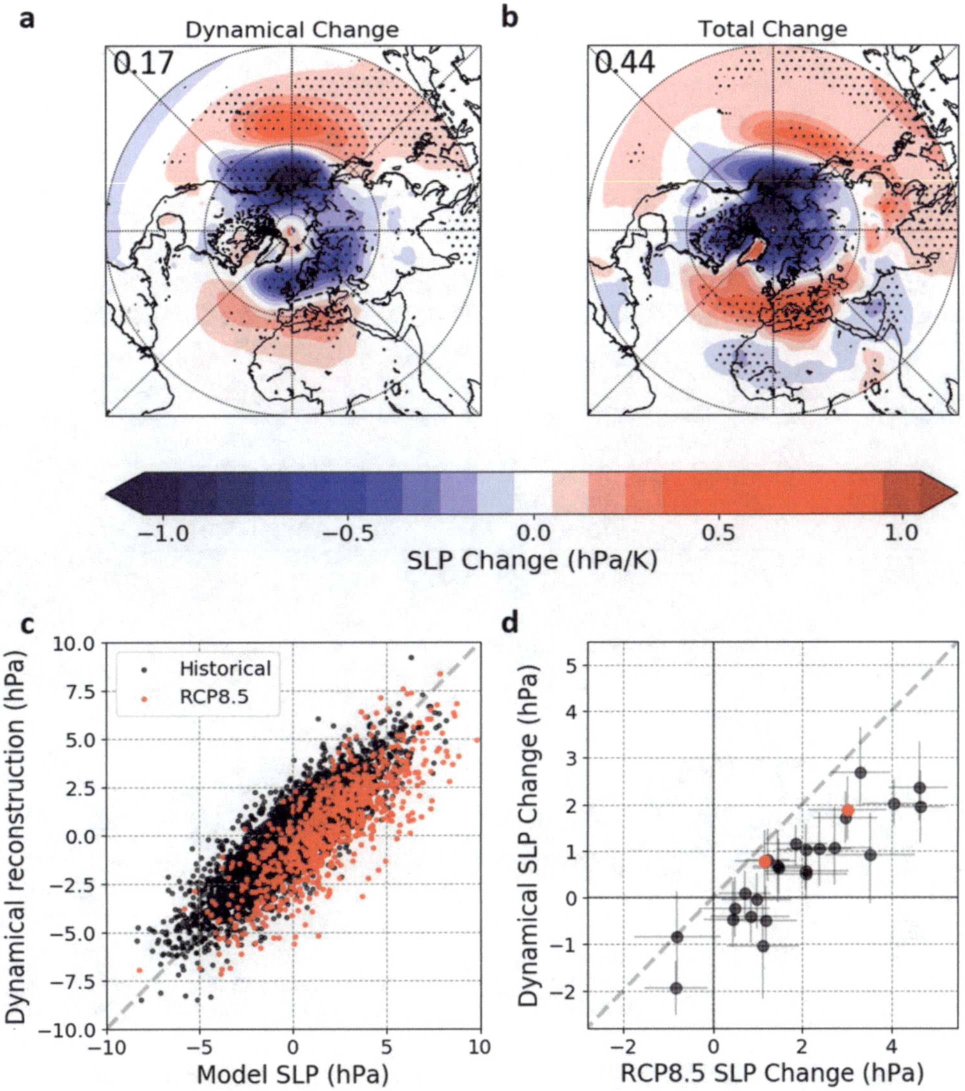

Figure 2.5: Northern Hemisphere CMIP5 multimodel mean change in DJF SLP (a) estimated by dynamical adjustment and (b) projected by CMIP5 models. Stippling shows agreement on the sign of the change by 80% ofmodels. Numbers indicate average value within the dash-outlined box. (c) Model vs dynamically reconstructed DJF Mediterranean (0°–30°E, 32°–48°N) SLP anomalies in historical (black dots) and RCP8.5 (red dots) simulations, forall 30 CMIP5 models. (d) Mediterranean (0°–30°E, 32°–48°N) CMIP5 intermodel change in DJF SLP estimated bythe dynamical adjustment model along with 90% confidence intervals. The two models used in Brogli et al. (2019) are highlighted with red dots.

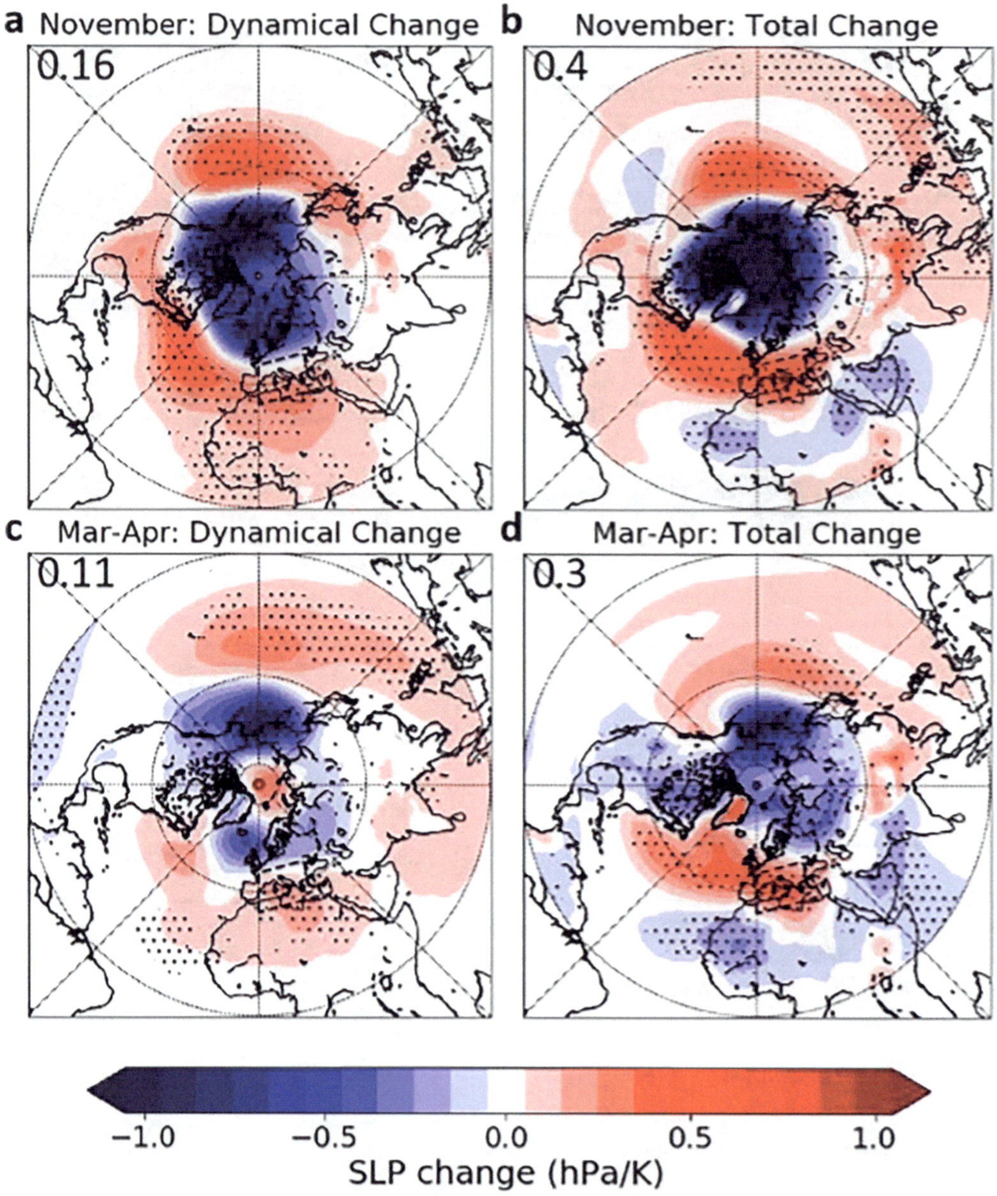

Figure 2.6: Northern Hemisphere SLP change: (a,c) CMIP5 multi-model mean for (a) November and (c) March-April, and (b,d) dynamical adjustment estimation for (b) November and (d) March- April. Numbers indicate average value within the box (dashed). Stippling shows agreement on the sign of the change by 80% of models.

explain only ~30%–40% of the whole signal (Figure 2.6).

Surface thermodynamical forcing

The geography of the Mediterranean Basin is unique, characterized by the existence of a large sea enclosed by continents on almost all sides. During winter, be- cause of water's larger thermal capacity, the Mediterranean Sea is on average warmer than the surrounding land. However, as a result of enhanced warming over land under climate change (Sutton et al. 2007; Byrne and O'Gorman 2018), future climate projections exhibit a robust and gradual relative cooling of the Mediterranean Sea with respect to surrounding land (Figure 2.7a). We define a simple sea–land temperature difference index as the difference between mean DJF temperatures over sea and over land in the Mediterranean region (5°W–38°E, 28°–48°N; Figure 2.7a). In historical CMIP5 simulations and ERA-Interim, that difference reaches about 5°C; by 2100, RCP8.5 simulations project a 0.5°C decrease, with little spread across models (Figure 2.7b). The magnitude of the projected index change is similar to its historical interannual standard deviation, suggesting a potentially important role of sea–land temperature change in shaping future regional circulation. Up to now, the projected decrease has been consistent with a range of observational datasets (Figure 2.7c). The impact of surface SST anomalies on regional circulations has been investigated, both from a theoretical perspective (Hoskins and Karoly 1981) and in model projections for the coming century (e.g., Gervais et al. 2019). At leading order, using geostrophic balance, a shallow relative cooling over the Mediterranean should result in anomalous surface anticyclonic circulation, since the land–sea warming contrast disappears in the mid- troposphere (Byrne and O'Gorman 2016). Linear theory similarly suggests a baroclinic response to a Mediterranean negative SST anomaly

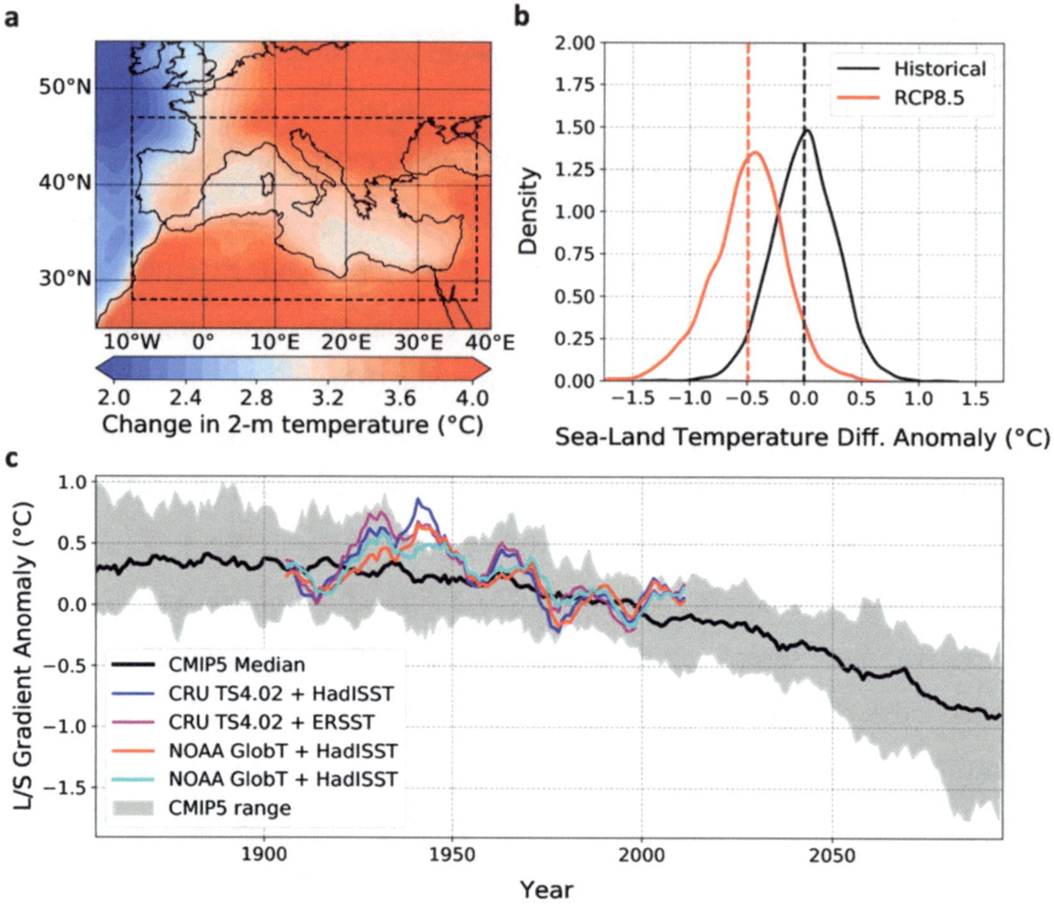

Figure 2.7 (a) CMIP5 multimodel DJF 2-m temperature change under RCP8.5 (2071–2100 minus 1976–2100). The dash-outlined box encompasses the area used to calculate the sea–land temperature index. (b) Distribution of historical (black) and RCP8.5 (red) DJF sea–land temperature index anomalies across CMIP5 models. (c) Five-year smoothed median value (black line) and range (gray shading) of CMIP5 model annual values (1850–2095) of DJF sea–land temperature index anomalies, alongside observations from CRU, NOAA Global Surface Temperature, ERSST, and HadISST. In (c) only, the temperature index is defined using SST over the sea and not2-m temperature, since the latter has limited availability over sea areas in observations.

characterized by a downstream SLP high (Hoskins and Karoly 1981). A similar result has been obtained while investigating the atmospheric response to the relative cooling of the NAWH in a realistic model (Gervais et al. 2019).

Model simulations

We investigate the impact of the land–sea warming contrast on Mediterranean circulation using regional climate simulations with the MIT Regional Climate Model (MRCM). MRCM is based on the Abdus Salam International Centre for Theoretical Physics Regional Climate Model, version 3 (RegCM3) (Pal et al. 2007), and includes several improvements (Winter et al. 2009; Gianotti et al. 2012; Gianotti and Eltahir 2014a,b) achieved through incorporation of new physical schemes or modification of original schemes. MRCM has been rigorously tested against observations in its ability to simulate key observed climate features across several regions (e.g., North America (Winter et al. 2009), West Africa (Im and Eltahir 2018b), the "Maritime Continent" (Im and Eltahir 2018a), and southwestern Asia (Pal and Eltahir 2016)).

We conduct two sets of simulations to assess the im- pact of the Mediterranean relative cooling over a do- main that encompasses the whole Mediterranean and Black Seas (Figure 2.8), at a horizontal resolution of 35 km and using 40 vertical levels. The first set is driven by the 1.5° × 1.5° 6-hourly ERA-Interim reanalysis (Dee et al. 2011) as lateral boundary conditions and 1° × 1° weekly NOAA OISST v2 (Reynolds et al. 2007) for the ocean surface, over the 1981–2011 period. In the second, we use the MPI-ESM-MR model (Zanchettin et al. 2013) under the RCP8.5 scenario as boundary conditions, over the 2070–2100 period. The MPI experiments are intended to assess the impact of a relative Mediterranean cooling in a climate change context as surrounding

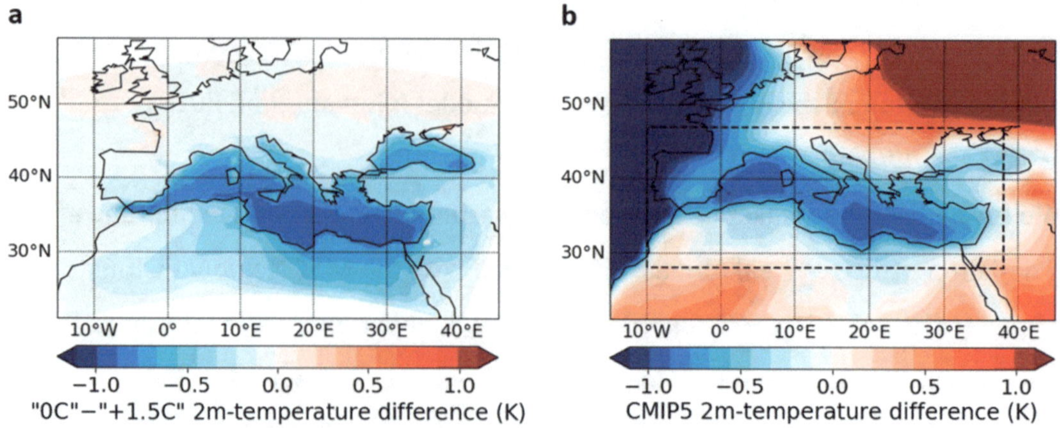

Figure 2.8 (a) Difference in DJF mean 2-m temperature between the 0°C and +1.5°C experiments (average of ERA and MPI simulations). (b) Change in DJF 2-m temperature under RCP8.5 (2071–2100 minus 1976–2005), fromwhich the average Mediterranean land warming (3.9 K) was subtracted to highlight the relative cooling over the sea.

continents warm. Still, any GCM presents a number of biases in its background climate, which may modulate the impact of the relative cooling. The ERA experiments thus allow us to assess the circulation impact of a relative cooling in the current Mediterranean climate with no land warming.

Each set consists of a reference simulation (called "0°C") and a perturbed simulation (called "+1.5°C"), in which Mediterranean SSTs are artificially increased by 1.5°C. The essence of the +1.5°C simulations is to force the Mediterranean to warm as much as if it were land. We choose an SST difference of 1.5°C because it translates in the simulations into a 0.5°–0.6°C change in the sea–land temperature index, consistent with the shift projected by CMIP5 models (Figures 2.7 and 2.9). The difference between the reference and the perturbed simulations (0°C minus +1.5°C) corresponds to the impact of the relative Mediterranean cooling on regional circulation. The first set of simulations is referred to as "ERA" ("ERA0C" and "ERA 1 1.58C"), and the second set is

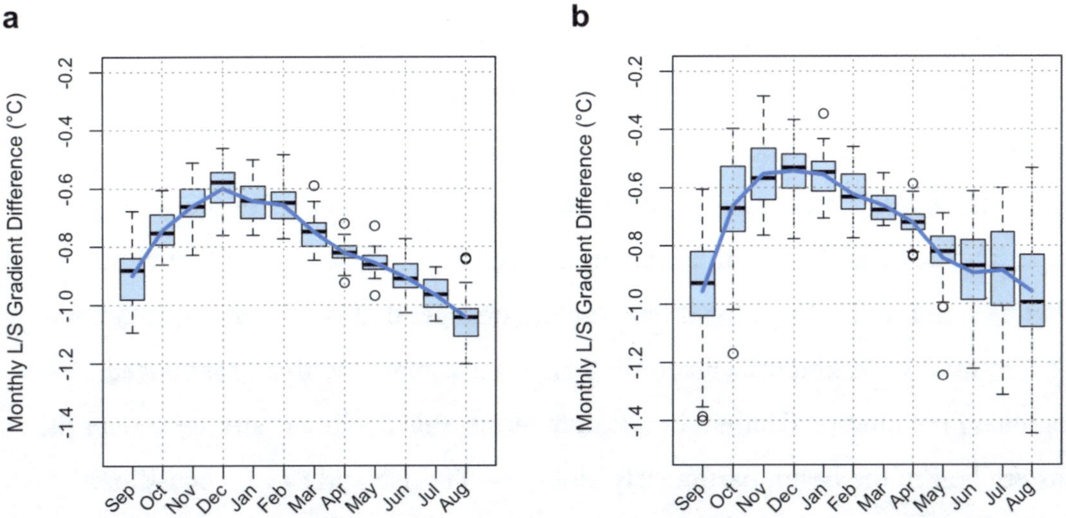

Figure 2.9: Annual cycles of sea-land temperature index difference between "0C" and "+1.5C" experiments, for (a) ERA and (b) MPI. Boxes represent the 75% inter-annual range, and whiskers the full range of inter-annual variability. "Sea" and "land" values are defined by the box on Figure 5-a.

referred to as "MPI" ("MPI0C" and "MPI + 1.5°C"). In all simulations, the first year is discarded as spinup. Long-term averages are computed over 1982–2011 for ERA experiments and 2071–2100 for MPI experiments.

Results

In comparison with the +1.5°C experiments, the 0°C experiments exhibit a roughly 0.6°C relative cooling of the sea–land temperature index (0.62°C for ERA and 0.57°C for MPI; Figure 2.9). The magnitude and spatial pattern of this relative cooling is similar to that projected by CMIP5 models (Figure 2.8). Figure 2.10 shows the difference between the two experiments (0°C minus 11.5°C) in mean DJF SLP, 850-hPa wind, and precipitation fields over our simulation domain. As compared with the +1.5°C simulations, the 0°C simulations exhibit a substantial anticyclonic flow in the boundary layer with a strong anomalous ridge generally located downstream of the SST anomaly. Both anticyclonic anomalies are accompanied by anomalous easterlies and northeasterlies over North Africa and the Middle East of about 0.5 m s^{-1} magnitude and a much weaker circulation response to the north. The pattern of the wind response south of 40°N is similar to that of the CMIP5 multimodel mean under RCP8.5, especially for the ERA experiments, even if of smaller magnitude (see Figure 2.1c, noting that wind change is normalized by global temperature change, that is, divided by a factor ≈ 4). The relative Mediterranean cooling is also associated with a pronounced regional precipitation decline. In absolute terms, this decline is strongest over the eastern half of the basin, particularly along the Turkish and Levant coastline. Precipitation is less affected west of 10°E, except along the Algerian coastline.

The main difference between the ERA and MPI simulations is the location of the anomalous high. It is centered south of Greece and peaks at about 1.1 hPa in ERA, whereas it is slightly weaker (≈1 hPa) and shifted southeast in the MPI experiment (Figures 2.10a,c). Averaging over the 0°–30°E, 32°–48°N domain, we find a 0.76-hPa SLP difference between ERA0C and ERA + 1.5°C and 0.62-hPa difference between MPI0C and MPI + 1.5°C. Additionally, the baroclinicity of the response is much more pronounced in the MPI simulations. However, the background climates in the two reference (0°C) simulations are different, notably the low-level zonal wind and static stability (Fig. S5 in the online supple- mental material). Those two fields are critical in deter- mining the atmospheric circulation response to shallow heating anomalies in the midlatitudes (Hoskins and Karoly 1981; Hall et al. 2001). The MPI reference state displays a slightly larger static stability over the Mediterranean and a stronger zonal wind field, both of which are consistent with a weaker circulation response to the imposed surface

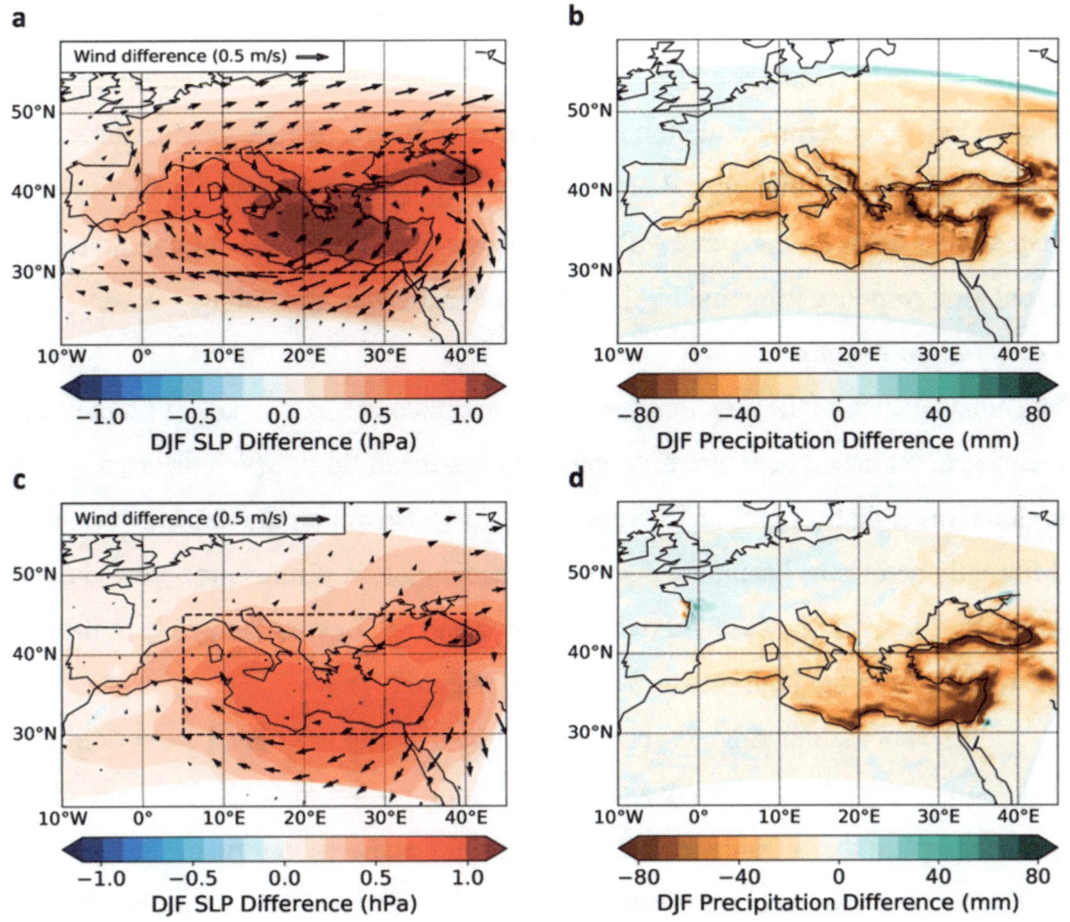

Figure 2.10: Difference in DJF (left) SLP and 850-hPa winds and (right) precipitation between the +1.5°C and 0°C experiments (0°C minus 1.5°C) in the (a),(b) ERA and (c),(d) MPI simulations.

temperature anomaly in the MPI experiments. A larger static stability limits the expansion of the heating anomaly in the lower troposphere, while the enhanced zonal wind reduces the circulation response required to compensate for the heating anomaly (Hoskins and Karoly 1981). Furthermore, the MPI simulations also include long-term circulation trends induced by greenhouse gas forcing. Despite this, the circulation and precipitation responses to the relative cooling are very similar between ERA and MPI. This points toward a robust anticyclonic circulation response triggered by the relative Mediterranean cooling.

Precipitation response

The consequence of the development of the anomalous Mediterranean ridge is an increase in mass and moisture divergence by the mean flow, which drives a regional precipitation decline (Seager et al. 2014). However, the impacts are felt differently across the basin: the decrease in precipitation is most pronounced over southern Turkey to the east and over Morocco and the Iberian Peninsula to the west, away from the high pressure anomaly (Figure 2.1).

The anomalous ridge is accompanied by a strong low- level wind field, that disrupts the relatively weak average circulation south of 40°N. Thermal balance implies that subsidence will develop on the eastern margin of the Mediterranean ridge to compensate for cold advection from anomalous northerlies (Seager et al. 2014), which will tend to suppress precipitation. This view is consistent with the particularly large anomalous downward motion centered over that region in future projections (Figure 2.11a).

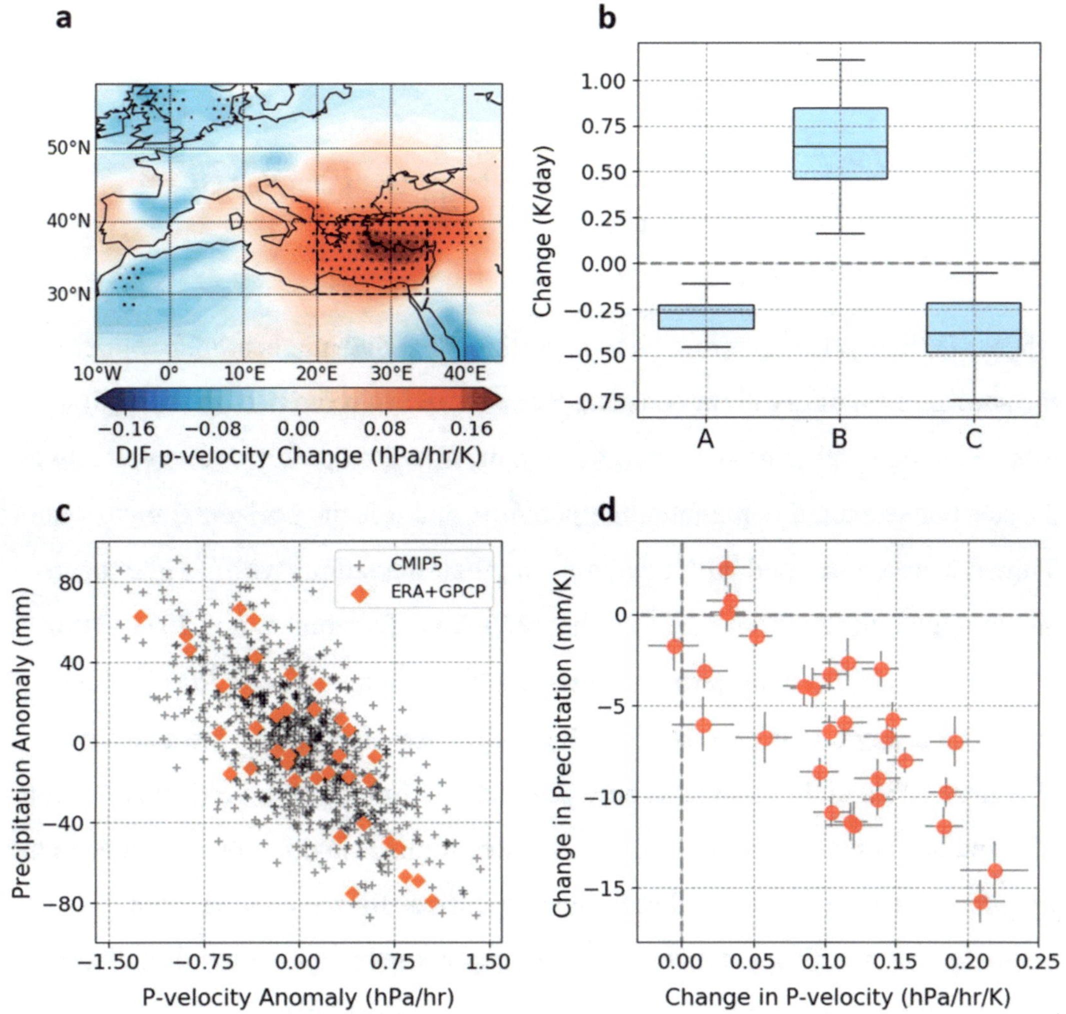

Figure 2.11: (a) CMIP5 multimodel mean change in DJF 500-hPa pressure velocity, under the RCP8.5 scenario. For each model, projections were normalized by its mean annual global temperature change. Dots indicate that more than 80% of models agree on the sign of the change. (b) Change in column-integrated energy budget terms across CMIP5 models, averaged over the eastern Mediterranean [see dash-outlined box in (a)] [A: diabatic warming, B: vertical advection, and C: horizontal advection; see Eq. 2.2]. (c) Eastern Mediterranean [20°–35°E, 30°–40°N; see box in (a)] DJF precipitation and pressure velocity anomalies in historical (1976–2005) CMIP5 models and the observed record (1979–2018: ERA-Interim pressure velocity and GPCP precipitation). (d) Change in eastern Mediterranean DJF precipitation and pressure velocity across CMIP5 models, normalized by each model's mean annual global temperature change; 95% confidence intervals are shown by light-gray bars.

The time-mean regional energy balance can be written in pressure coordinates as (Rodwell and Hoskins 1996)

$$\overline{\frac{\partial T}{\partial t}} = \underbrace{\frac{\bar{Q}}{c_p}}_{\text{A}} - \underbrace{\left(\frac{p}{p_0}\right)^K \bar{\omega}\frac{\partial \bar{\theta}}{\partial p}}_{\text{B}} - \underbrace{\bar{\mathbf{u}} \cdot \nabla_p \bar{T}}_{\text{C}} - \left(\frac{p}{p_0}\right)^K \frac{\partial}{\partial p}\overline{(\omega'\theta')} - \nabla_p \cdot \overline{(\mathbf{u}'\mathrm{T}')} \qquad (2.2)$$

where t is time, p is pressure, T is temperature, Q is diabatic heating/cooling, c_p is the specific heat of dry air at constant pressure, p_0 is a reference pressure (1000 hPa), R is the specific gas constant for dry air, $K = R/c_p$, ω is pressure velocity in pascals per second, θ is potential temperature, and **u** is the horizontal wind vector. Time means are denoted with overbars, and their deviations with $'$. Time means are computed in CMIP5 models for the 1976–2005 (reference) and 2071–2100 (future) periods. For long-term averages, the term on the left $\overline{(\partial T/\partial t)}$ is about zero. The last two terms on the right- hand side (transient terms) are also small (Figure 2.12), and the balance is between diabatic forcing (A), vertical advection (B), and horizontal advection (C). Those three terms also dominate changes in the long-term balance. Downward motion compensates for the cooling from anomalous northerlies; yet, because it also suppresses precipitation, it indirectly leads to diabatic cooling of the air column (less vertical latent heat flux) and therefore reinforces itself. This likely explains why the cooling from horizontal advection is only about two-thirds of the change in adiabatic warming (Figure 2.11b). Models correctly reproduce the inter- annual link between winter eastern Mediterranean\ precipitation and vertical velocity anomalies (Figure 2.11c).

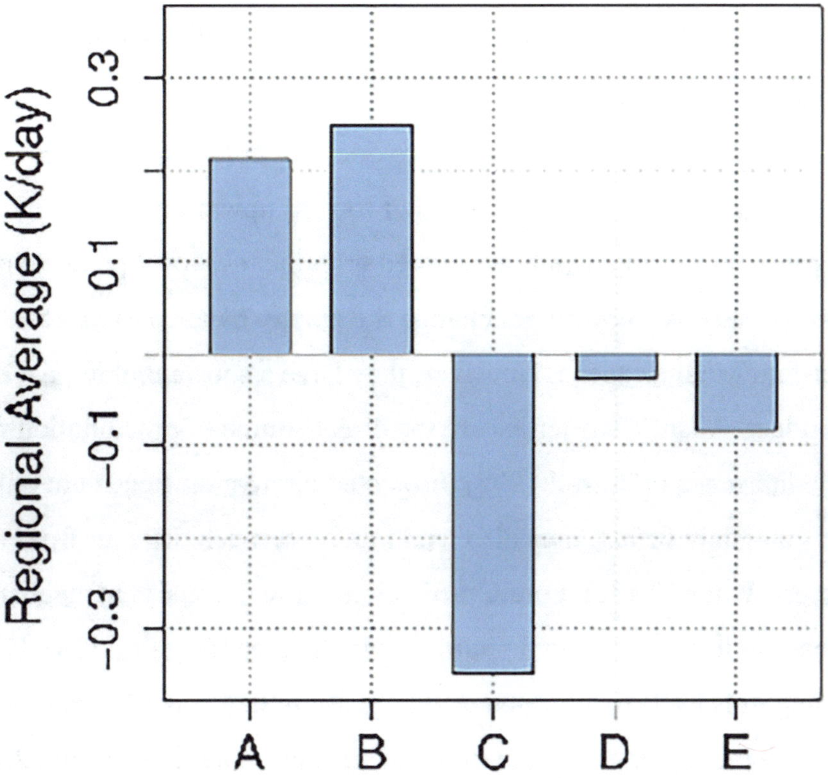

Figure 2.12: Energy budget equation terms for the Eastern Mediterranean (20° - 35°E/30° - 40°N), estimated from ERA-Interim 6-hourly data (1980-2018).

$$A = \frac{\bar{Q}}{C_p},$$

$$B = -\left(\frac{p}{p_0}\right)^K \bar{\omega}\frac{\partial\bar{\theta}}{\partial p},$$

$$C = -\bar{\mathbf{u}} \cdot \nabla_p \bar{T},$$

$$D = -\left(\frac{p}{p_0}\right)^K \frac{\partial}{\partial p}\overline{(\omega'\theta')},$$

$$E = -\nabla_p \cdot \overline{(\mathbf{u}'T')}.$$

D and E correspond to transient terms.

Across models, the change in seasonal precipitation is well correlated with that in pressure velocity (Figure 2.11d).

To the west of the basin, this picture does not hold. There, while precipitation still declines, the trend is rather toward upward motion (Figure 2.11a). Because anomalous winds associated with the low-level pressure high are predominantly easterly, they do not change the energy balance as much as they do over the eastern Mediterranean. However, they have a substantial impact on the moisture budget. Atlantic westerlies are the direct source of precipitation for the western Mediterranean (Hurrell 1995). Projected circulation trends not only act to weaken this moisture influx, they also tend to advect much drier air from the Sahara Desert (Figure 2.13c). Future projections show a consistent stagnation or weak increase in low-level specific humidity in the area (Figure 2.13a). Combined with warming, this leads to a pronounced increase in dewpoint depression, well correlated with the regional precipitation decline across models (Figure 2.13b).

In addition, as noted above, the relative cooling of Mediterranean SSTs may act to decrease precipitation not only through its effect on regional circulation but also by limiting the increase in evaporation and therefore moisture fluxes toward land. This might be an important factor for the eastern Mediterranean, which receives much of its winter precipitation from Mediterranean cyclones (Zappa et al. 2015a) (Figure 2.10b,d).

While these elements seem to explain precipitation trends in GCMs, it is important to remember that both the eastern and western Mediterranean exhibit in reality complex orographic features, which play an important role in regional precipitation patterns. Their interaction with large-scale circulation trends may

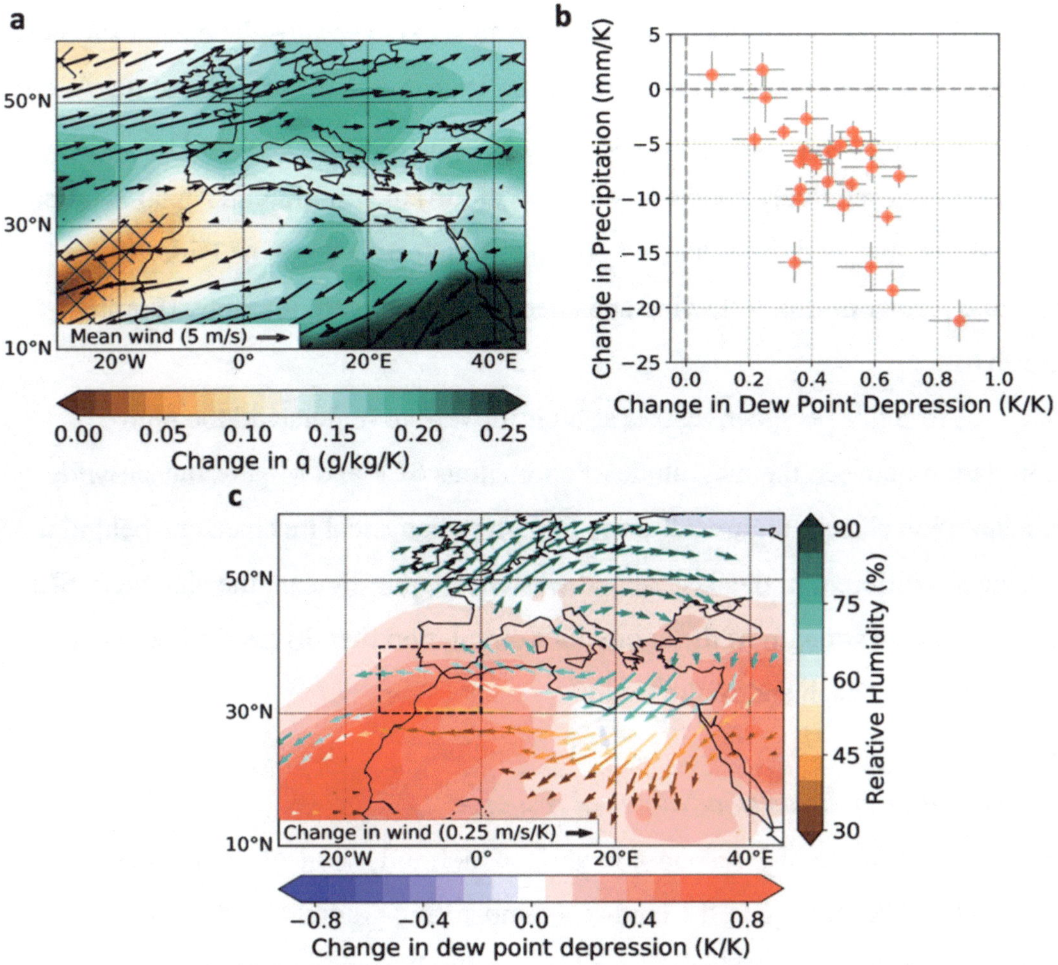

Figure 2.13: CMIP5 multimodel mean change in DJF 925–850-hPa averaged (a) specific humidity and (c) dewpoint depression, normalized by each model's annual global-mean temperature change. Arrows indicate DJF historical [in (a)] and change in [in (c)] 925–850-hPa averaged winds. Wind arrows in (c) are colored according to mean historical relative humidity at each point and are shown only where 80% of models agree on the sign of the change.

(b) Western Mediterranean [15°W–0°, 30°–40°N; dash-outlined box in (c)] DJF precipitation and dewpoint de- pression change across CMIP5 models, normalized by each model's mean annual global temperature change; 95% confidence intervals are shown by light-gray bars.

also impact future precipitation changes. Over southern Turkey, anomalous northerlies forced down the Anatolian plateau will cause subsidence and suppress precipitation. A similar phenomenon can be expected to occur north of the Atlas Range in Morocco and Algeria. These features are not well resolved by GCMs; already at our relatively coarse resolution (35 km) the orographic signal is much more detectable, and the enhanced precipitation decline south of Turkey may indicate further drying following subsidence induced by topography (Figure 2.10b,d).

From this perspective, it is straightforward to understand the high correlation between the magnitude of anomalous SLP and Mediterranean-wide precipitation change Figure 2.3 Even though the physical mechanisms behind the drying are different in the eastern and western Mediterranean, the change in SLP determines the strength of the anomalous circulation that drives the decline in precipitation at each end of the basin.

Discussion

Uncertainties in the thermodynamical response

The estimate of the SLP response to surface thermodynamical forcing is based on a roughly 0.5°C change in the land–sea temperature gradient, with limited variance across models. That change is in no small part determined by Mediterranean SST trends under continued greenhouse gas forcing, which in turn depend, among others, on Mediterranean Sea circulation. The latter, however, is not well resolved in GCMs (Planton et al. 2012). Reasons are to be found among the region's complex geography, its small size relative to model resolution and the very small extent (15 km) of the Gibraltar Strait, which plays a major role in the mass and energy balance of the Mediterranean (Sannino et al. 2002). Energy exchange with the Atlantic Ocean at Gibraltar must be parameterized in GCMs, which of course leaves room for substantial errors in long-term SST trends. In some models, the Mediterranean is even represented as a closed sea, which may tend to overestimate the SST response, with implications on the regional circulation change.

Despite obvious differences, the impact of the Mediterranean relative cooling on regional circulation and precipitation is relatively similar between the ERA and MPI experiments: a general increase in SLP of about 0.6 hPa, peaking over the eastern Mediterranean at around 1 hPa. This suggests that the response to relative cooling may also be similar when using other CMIP5 models as lateral boundary conditions.

Combining mechanisms and other explanations

Future Mediterranean circulation and hydroclimate trends in the CMIP5 multimodel mean appear to be primarily determined by large-scale circulation changes and regional thermodynamical forcing. Both mechanisms lead to higher pressure over the Mediterranean and can explain why the region is picked out as a prominent climate change hot spot. While other factors, like soil moisture feedback or stratospheric vortex strength (Zappa and Shepherd 2017; Simpson et al. 2018) can never be totally excluded, they seem unlikely to contribute much to the multimodel mean. In individual models, they may certainly play a large role, thus affecting intermodel spread.

The "dynamical" SLP response is higher over the western Mediterranean, which can be seen as an eastward shift of the North Atlantic subtropical high, consistent with the eastward shift of the stationary waves themselves (Simpson et al. 2016). Contrarily, the SLP anomaly due to relative Mediterranean cooling

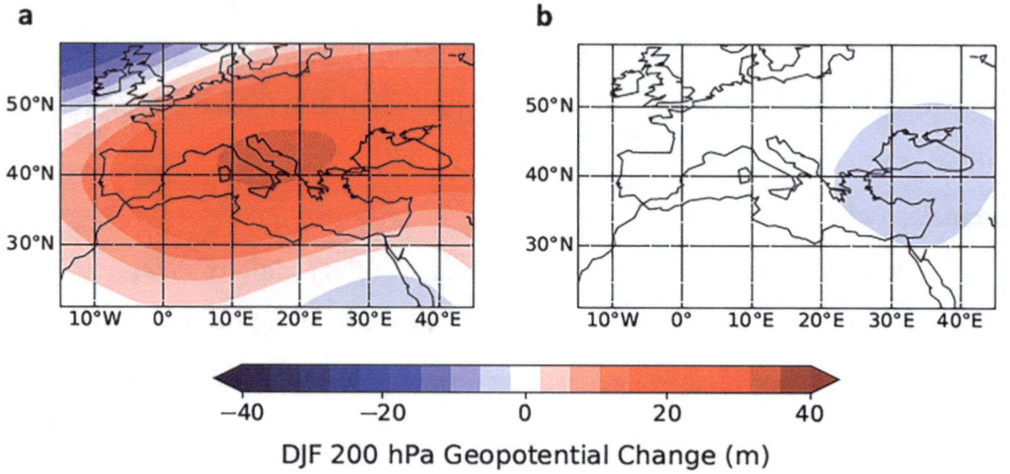

Figure 2.14: (a) CMIP5 multi-model mean Northern Hemisphere 200 hPa geopotential height change under RCP8.5 (2071-2100 minus 1976-2005). The zonal-mean change is subtracted at each latitude to highlight zonal variability. (b) Difference in DJF 200 hPa geopotential between the "0C" and "1.5C" experiments (average of ERA and MPI simulations).

peaks downstream of the anomaly, that is, over the eastern Mediterranean. Although assessed within different modeling frameworks, our approach suggests that the two mechanisms may individually contribute somewhat equally to the Mediterranean SLP increase (about +0.6–0.7 hPa, or 35%–40% of the mean projected response), and therefore to the precipitation decline. The dynamical SLP anomaly nevertheless clearly dominates in a number of models, in addition to being responsible for most of the intermodel spread. The strong geopotential ridge projected at upper levels above the Mediterranean by CMIP5 models (Seager et al. 2014) is also clearly associated with the planetary wave shift, not with the surface relative cooling, whose effect is limited to the lower troposphere (Figure 2.14).

A mutual reinforcement of the low-level anticyclonic circulation triggered by each forcing is, however, not to be excluded. Indeed, adding the SLP trends induced by our two mechanisms accounts for roughly 80% of the model-mean projected change, leaving a sizeable part unexplained. This unexplained fraction is even larger in several models. The discrepancy might result from nonlinear superposition (through self-reinforcement of the anomalies at low levels) or via the effects of surface friction, which also impacts the SLP response. In that regard, variability in regional orography and parameterization of friction may play an important role in explaining intermodel spread.

One other prominent feature of the midlatitude circulation change is an increase of SLP in the North Atlantic, between 45°W and 0° (Figure 2.1a). We find that part of this increase, along the 40°N parallel, where CMIP5 SLP trends are robust, may also be a response to upper-level circulation changes (Figure 2.5a). North of there, the SLP change, though less robust, may be the down-stream response to the NAWH, an area of relative SST cooling (Gervais et al. 2019), though this remains far from certain. Despite weaker intermodel agreement in this region during winter, a downstream increase in SLP extending to the Mediterranean is possible. Though it would not explain the robustness of the Mediterranean hot spot, it may combine with the robust anomalous positive SLP from our two proposed mechanisms. Model agreement in this region is stronger in spring and autumn, and this phenomenon could therefore also enhance the Mediterranean SLP response in those seasons, especially given the weaker projected upper- level flow changes in early spring and late autumn (Figure 2.6). Simulation results of Gervais et al. (2019) suggest that the impact is small but not zero (their Figures 3 and 5). They also used only one model. Analyzing, within a coherent modeling frame- work, the contributions of upper-level flow changes, the Mediterranean relative cooling, and the NAWH to future Mediterranean circulation trends may shed more light on this.

Consistency of mechanisms with other studies

We end by discussing the consistency of our results with those of previous studies. Among the various mechanisms proposed to explain the anomalous ridge, discussed in the introduction, we show in this paper that changes in stationary wave structure discussed by Simpson et al. (2016) play a crucial role, both for the model mean and intermodel spread (Figure 2.5). Consequently, poleward shifts of the Hadley circulation and storm tracks, already shown to be limited in the Northern Hemisphere during winter, probably have limited direct influence. Recently, Brogli et al. (2019) concluded that large-scale circulation changes were the main driver the Mediterranean precipitation decline, and that regional effects were minimal. Their simulations that focus on the relative cooling of the Mediterranean also find a decrease in precipitation in the eastern Mediterranean compared to a scenario where SSTs warmed as much as the land (their Figure 4). Yet, their conclusions were based on two models (MPI-ESM-LR and HadGEM2-ES) in which thc contribution of upper- tropospheric forcing happens to dominate (Figure 2.5d). As we show here, however, large-scale dynamics have a wide range of effects across models; although they account for much of the spread in the precipitation response, large-scale circulation trends are not sufficient to fully explain regional projections.

In terms of precipitation response, our results agree with the perspective offered by Seager et al. (2014). In CMIP5 models, the Mediterranean drying is connected to changes in moisture convergence by the mean flow, but the drivers behind those changes vary from east to west. To the west, moisture convergence declines due to a weakening of the Atlantic moisture supply through increased advection of dry air from the Sahara. To the east, large-scale subsidence is the main cause of the precipitation decrease. In reality, however, precipitation is strongly linked to prominent topographic features that are, to a large extent, not resolved by GCMs. Orographic lifting and the interaction of mountains with large-scale circulation changes predicted by CMIP5 models will im- pact precipitation projections.

Our results also shed some light on the conclusions of Zappa et al. (2015a), who found the eastern Mediterranean precipitation decline to be associated with large reductions in cyclone frequency and intensity. We find that regional circulation and its changes control precipitation variability and change in the basin (Figures 2.3, 2.11, and 2.13) and therefore should control cyclonic activity, which is responsible for most of winter precipitation (Zappa et al. 2015a). The decrease in mean storm precipitation and dynamical weakening of cy- clones they identified is consistent with the substantial anomalous regional downward motion over the region (Figure 2.11a). In addition, anomalous northerlies advecting drier higher-latitude air and the relative Mediterranean cooling both act to constrain humidity trends, which likely cause the projected local minimum in specific humidity change near Cyprus (Figure 2.13a). The weakening of the storm track is consistent with a weakening of the meridional temperature gradient between the Mediterranean Sea and the continent to its north, and the reduction in baroclinicity it may lead to. Future research should focus on understanding how the large-scale drivers we identified control storm genesis in the Mediterranean.

Conclusions

In conclusion, our results offer a coherent and mechanistic picture of Mediterranean winter climate projections that is consistent with their robustness. It involves sound physical mechanisms and characteristics of the mean circulation: the shift in wavelengths of stationary waves, enhanced warming over land compared to oceans, and the baroclinic response to surface temperature anomalies. While much work remains to be done to understand the variability in SLP and precipitation change patterns across models, we find that the unique combination of the large-scale "dynamic" response and the independent, regional-scale "thermodynamic" response explain the robust regional maximization of wintertime climate change over the Mediterranean, making this region a major climate change hot spot for the coming decades. The contributions from each response appear comparable, though a nonlinear superposition may contribute to enhancing the response even more in certain models. Robust physical mechanisms involving local circulation account for the coherence of regional precipitation change and a maximized response at each end of the basin. Complex topography in those two regions indicates, however, that detailed regional simulations are required for improved future climate projections.

Acknowledgments

We acknowledge the World Climate Research Programme's Working Group on Coupled Modelling, which is responsible for CMIP, and we thank the climate modeling groups for producing and making available their model output. This work was made possible through funding from the Office Chérifien des Phosphates (OCP) through Université Mohamed VI Polytechnique, Morocco. The authors thank R. A. Plumb and P. O'Gorman for their remarks.

Name	Resolution: Atmosphere	Resolution: Ocean	Main reference
ACCESS1-0	144 × 192	300 × 360	Bi et al. (2013)
ACCESS1-3	144 × 192	300 × 360	
BCC-CSM1-1-m	160 × 320	232 × 360	Xin et al. (2013)
BNU-ESM	64 × 128	–	Ji et al. (2014)
CanESM2	64 × 128	192 × 256	Arora et al. (2011)
CCSM4	192 × 288	384 × 320	Gent et al. (2011)
CESM1-CAM5	192 × 288	384 × 320	Neale et al. (2010)
CMCC-CESM	48 × 96	149 × 182	
CMCC-CM	240 × 480	149 × 182	Scoccimarro et al. (2011)
CMCC-CMS	96 × 192	149 × 182	
CNRM-CM5	128 × 256	292 × 362	Voldoire et al. (2013)
CSIRO-Mk3-6-0	96 × 192	189 × 192	Rotstayn et al. (2010)
FGOALS-g2	60 × 128	360 × 196	Li et al. (2013)
GFDL-ESM2G	90 × 144	210 × 360	Dunne et al. (2012)
GFDL-ESM2M	90 × 144	200 × 360	
GISS-E2-R	90 × 144	144 × 90	Miller et al. (2014)
GISS-E2-R-CC	90 × 144	144 × 90	
HadGEM2-AO	144 × 192	216 × 360	Collins et al. (2011)
HadGEM2-ES	144 × 192	216 × 360	
INM-CM4	120 × 180	340 × 360	Volodin et al. (2010)
IPSL-CM54-LR	96 × 96	149 × 182	Marti et al. (2010)
IPSL-CM54-MR	143 × 144	149 × 182	
MIROC-ESM	64 × 128	192 × 256	Watanabe et al. (2010)
MIROC-ESM-CHEM	64 × 128	192 × 256	
MPI-ESM-LR	96 × 192	220 × 256	Zanchettin et al. (2013)
MPI-ESM-MR	96 × 192	404 × 802	
MRI-CGCM3	160 × 320	368 × 360	Yukimoto et al. (2012)
NorESM1-M	96 × 144	384 × 320	Bentsen et al. (2012)
NorESM1-ME	96 × 144	384 × 320	

Table 2.1: List of CMIP5 models used in this study. Resolution is in latitude by longitude. The selection was made based on available output at the desired time resolution and scenarios.

A version of this chapter first appeared as Tuel, A. and Eltahir, E., 2020. Why Is the Mediterranean a Climate Change Hot Spot?. Journal of Climate, 33(14), pp.5829-5843. (Original article under AMS copyright)

Chapter 3: Elements of the Dynamical Response to Climate Change over the Mediterranean

Alexandre Tuel
Paul A. O'Gorman
Elfatih A. B. Eltahir

The Mediterranean Basin (30°–45°N, 10°W–40°E) has stood out in successive generations of climate models as one of the few regions where circulation projections are particularly robust. In particular, a large majority of models in both CMIP3 and CMIP5 agree on the development of an anomalous pressure high during winter (Giorgi and Lionello 2008; Seager et al. 2014), whose magnitude largely surpasses that of changes elsewhere in the Northern Hemisphere (Figure 3.1). Changes in regional winter circulation have long been suspected to drive the projected decrease in the frequency of Mediterranean cyclones and the concurrent regional precipitation decline (Zappa and Shepherd 2017). North African wind anomalies were shown to be strongly related to regional precipitation anomalies and explained most of the latter's scatter across CMIP5 models (Zappa et al. 2015). Regional circulation trends were subsequently linked to large-scale

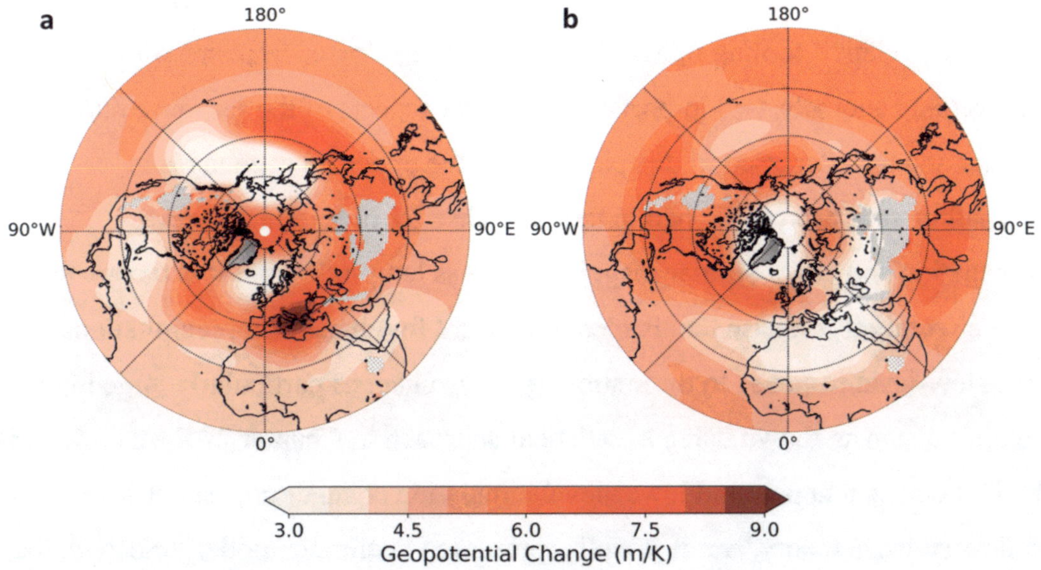

Figure 3.1: CMIP5 multimodel mean change (2071–2100 minus 1976–2005) in (a) DJF and (b) JJA 850-hPa geo- potential height, normalized by each model's global-mean projected temperature change. Gray hatching indicates areas where the 850-hPa level is below the surface.

dynamics (Seager et al. 2019; Brogli et al. 2019; and see Chapter 2 of this book) and the land-sea warming contrast (Brogli et al. 2019; see also Chapter 2). We explained in Chapter 2 the development of the anomalous Mediterranean high as the combined response to shifts in the large-scale structure of the mean wintertime circulation and to surface temperature trends. A robust eastward shift in winter stationary waves (Simpson et al. 2016) pushes the Atlantic Ocean subtropical high closer to the Mediterranean Sea while the relative cooling of Mediterranean sea surface temperatures (SSTs) in comparison with the surrounding land triggers higher pressure over the eastern part of the basin. The link between circulation and precipitation trends was discussed in Chapter 2, who showed how the strong wind anomalies associated with the increase in surface pressure suppressed precipitation at each end of the basin.

Although we now have a good understanding that both shifts in stationary waves and relative cooling of SSTs contribute to the lower-tropospheric anticyclonic anomaly, two important dynamical aspects of the response remain unclear and are addressed in this paper. First, the upper-level changes in circulation have been explained in terms of changes in stationary wave propagation and are of similar magnitude across longitudes, but it is the lower-level circulation response that is most important for changes in precipitation. Why is the low-level response to the stationary wave changes particularly large in the Mediterranean as shown using a statistical approach in Chapter 2? Furthermore, is the link between upper- and lower-level circulation changes consistent with stationary-wave theory, and is it well represented in climate models? Second, the relative cooling of the Mediterranean SSTs occurs both in summer and winter, but the anomalous high pressure only occurs in winter. Why does the response to relative cooling of the Mediterranean Sea lead to a large dynamical response in winter and not in summer?

Data

As estimate of the current climate, we consider the ERA-Interim reanalysis (Dee et al. 2011) over the 1979–2018 period. ERA- Interim is based on an atmospheric model and reanalysis system with a horizontal grid spacing of about 79 km. Here, we perform our calculations with data archived on a regular 1° grid with 17 pressure levels at monthly resolution. We also analyze global climate model (GCM) output from 28 models of phase 5 of the Coupled Model Intercomparison Project (CMIP5; Taylor et al. 2012) under the historical and RCP8.5 scenarios, and 11 models from the Atmospheric Model Intercomparison Project (AMIP; Gates et al. 1999), under the "amip" and "amipFuture" scenarios. In amip, the atmosphere is forced by observed time-varying SSTs and sea ice at present-day atmospheric composition, while a patterned SST anomaly is added in the amipFuture runs (sea ice and CO_2 being kept the same). The SST pattern is based on the CMIP3 multimodel-mean SST response to a quadrupling of CO2, where individual CMIP3 model responses are normalized to have a global- mean 4K warming (see details at https://www.earthsystemcog.org/ projects/cfmip/cfmip2-cmip5). The list of models is given in Table 2.1 in the previous chapter. For the CMIP models, we use as reference the 1976–2005 period in the historical runs, and the 2071–2100 period under RCP8.5 for long-term projections. For AMIP, the amip and amipFuture runs both cover the same period, usually 1979–2008, but with some variability across models (see Table S2). For runs longer than 30 years, we keep only the last 30 years for analysis; otherwise, all available years are retained. The amip runs are used as reference. Our analysis also relies on output from the regional climate simulations developed and detailed in Chapter 2 using the MIT Regional Climate Model. We ran two sets of simulations over a domain covering the Mediterranean Sea and most of Europe: one forced with ERA-Interim data as boundary conditions over the 1981–2011 period ("ERA"), and one forced with output from the MPI-ESM-MR model (Zanchettin et al. 2013) under the RCP8.5 scenario between 2070 and 2100 ("MPI"). Each set comprises a reference (unperturbed) run (designated by "0°C"), and a perturbed run in which Mediterranean SSTs are uniformly warmed by 1.5°C year-round (designated by "+1.5°C"). The +1.5°C perturbed experiment is designed to "cancel out" the future wintertime relative cooling of the Mediterranean with respect to the surrounding land projected by CMIP5 models (see Chapter 2). Thus, this framework allows one to estimate the impact of the relative SST cooling on the regional climate by taking the difference between the +1.5°C and 0°C simulations.

Large-scale dynamics

Changes in stationary planetary waves

Wintertime midlatitude flow in the Northern Hemisphere is strongly modulated by atmospheric stationary waves, which arise from orographic and atmospheric diabatic forcing, land–sea thermal contrasts, as well as transient eddy heat and momentum fluxes (Hoskins and Karoly 1981; Held et al. 2002). Stationary waves exhibit wavelengths ranging roughly from a few thousand to twenty thousand kilometers, and generate large-scale variations in the atmospheric circulation, leading to large longitudinal asymmetries in the distribution of temperature and precipitation. Global climate models robustly agree on the pattern of change in upper- tropospheric wintertime circulation in the Northern Hemisphere (Simpson et al. 2016; Wills et al. 2019), consisting of a strengthened jet stream and alternating cyclonic and anticyclonic circulation anomalies between 20° and 60°N. The acceleration of the jet stream follows from the thermal wind balance response to the increase in the equator-to-pole temperature gradient at upper levels, itself a consequence of enhanced warming in the middle and upper tropical troposphere and less warming in the upper troposphere at higher latitudes (Rind 2008). Given that the structure and amplitude of stationary waves is very dependent on the strength of the jet stream, changes in zonal and meridional winds are strongly related. The pattern of change in upper-tropospheric meridional winds can be understood from simple theoretical arguments (Simpson et al. 2016). Linear theory of barotropic Rossby waves on a zonal-mean zonal flow u_0 away from sources states that the total wavenumber for stationary waves is given by

$$K = \sqrt{k^2 + l^2} = \sqrt{\frac{\beta - \frac{\partial^2 u_0}{\partial y^2}}{u_0}}, \qquad (3.1)$$

where k and l are the zonal and meridional wavenumber, respectively [e.g., Hoskins and Ambrizzi 1993]. In the current climate, stationary waves with nondimensional zonal wave- numbers smaller than 7 are largely prevalent (Simpson et al. 2016). As u_0 increases, K must decrease. Assuming unchanged l, the zonal wavenumber must also decrease, which leads to an increase in the wavelength of stationary waves. Because the wave sources are mainly tied to Earth's topography and are therefore largely fixed in space, the change in wavenumber translates into a phase change downstream from sources, where stationary waves extend further eastward in future climate projections (Figure 3.2a).

Upper-tropospheric meridional wind changes are somewhat stronger and more robust over North America than the Mediterranean, yet overall the magnitude of meridional wind changes is rather uniform across longitudes (Figure 3.2b). A direct consequence of the eastward shift is the development of an anomalous upper-level anticyclonic circulation over the Mediterranean. This anomalous pattern is also expected at lower levels due to the generally equivalent barotropic nature of winter stationary waves (Held et al. 1985, 2002). The wavelike pattern is indeed still present at the 850-hPa level but does not follow a simple dampening of the 300-hPa pattern. The largest circulation changes are now found over the Mediterranean and off the western coast of North America, whereas meridional wind anomalies are quite weak over the tropical Atlantic, the

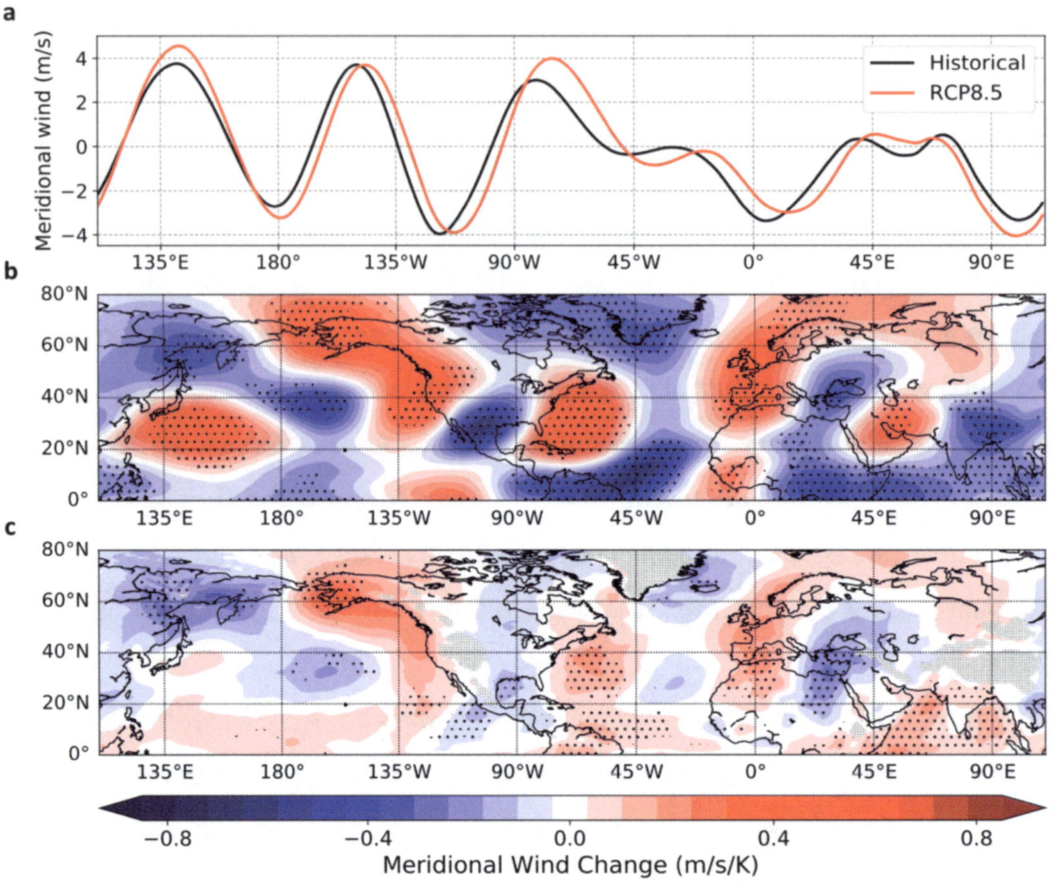

Figure 3.2: (a) CMIP5 multimodel mean 300-hPa meridional winds in DJF in the historical and RCP8.5 scenarios, averaged between 208 and 508N. Also shown are CMIP5 multimodel mean projected change in DJF meridional winds at (b) 300 and (c) 850 hPa. Gray hatching in (c) indicates areas where the 850-hPa level is below the surface. For each model, projections were normalized by its mean annual global temperature change. Stippling indicates that more than 80% of models agree on the sign of the change.

Gulf of Mexico, and the Arabian Peninsula. While there is clearly an overall equivalent barotropic structure over the North Atlantic, Pacific and the Mediterranean, the pattern appears more baroclinic over Asia and North America.

Significant longitudinal variations in the vertical structure of Northern Hemisphere winter circulation have been documented by Blackmon et al. (1979). They found that over the central and eastern parts of the Atlantic and Pacific oceans, and Western Europe, the vertical mode is highly barotropic with strong vertical coherence. On the other hand, North America and Asia are characterized by a more baroclinic structure with little coherence between low- and upper-level fluctuations. It follows from this observation that lower-tropospheric flow over the large continental masses should be much less sensitive to stationary circulation changes than over oceans, which is consistent with the comparison between 300- and 850-hPa meridional wind changes. To find the physical cause of this empirical finding, we turn to linear theory of midlatitude dynamics.

The external mode

For atmospheric conditions typically prevailing in the midlatitudes, there exists a single trapped vertical mode for Rossby waves propagating on a mean westerly zonal flow with vertical shear, referred to as the external mode (Held et al. 1985). Since this mode therefore connects the upper- and lower-tropospheric circulation, differences in the surface response to similar high-level circulation anomalies should be partly related to variability in the external mode itself. A lack of vertical coherence of the stationary circulation could also locally relate to vertically propagating Rossby waves over regions of topography, but here we focus on the structure of the external mode since it should be an important factor over all land regions as well as over ocean.

The vertical structure of the external mode can be calculated by solving the stationary quasigeostrophic potential vorticity and thermodynamic equations on a beta-plane of infinite zonal extent, linearized about a mean zonal flow $\bar{u}(p)$ between the surface and the 50-hPa level:

$$\frac{\partial q}{\partial t} + \bar{u}\frac{\partial q}{\partial x} + \frac{\partial \psi}{\partial x}\left[\beta - \frac{\partial}{\partial p}\left(\frac{f^2}{\sigma}\frac{d\bar{u}}{dp}\right)\right] = 0, \qquad (3.2)$$

where

$$q = \Delta\psi + \frac{\partial}{\partial p}\left(\frac{f^2}{\sigma}\frac{\partial \psi}{\partial p}\right)$$

is the anomalous quasigeostrophic potential vorticity, ψ is the anomalous streamfunction, f is the Coriolis parameter, β is its meridional gradient, $\omega = Dp/Dt$, and

$$\sigma = -\frac{RT}{p}\frac{d\log(\theta)}{dp}$$

is the static stability parameter. At the surface and at 50 hPa, we obtain boundary conditions from the thermodynamic equation

$$\left(\frac{\partial}{\partial t} + \bar{u}\frac{\partial}{\partial x}\right)\frac{\partial \psi}{\partial p} - \frac{\partial \psi}{\partial x}\frac{d\bar{u}}{dp} = -\frac{\sigma\omega}{f} \qquad (3.3)$$

by enforcing $\omega = 0$. In looking for stationary solutions of the form $\psi(x, y, p) = \mathrm{Re}\left[\hat{\psi}(p)e^{i(kx+ly)}\right]$, Equation (3.2) becomes

$$\frac{1}{\bar{u}}(\bar{u}\Gamma + \beta - \Gamma\bar{u})\hat{\psi} = K^2\hat{\psi}, \qquad (3.4)$$

where

$$\Gamma = \frac{\partial}{\partial p}\left(\frac{f^2}{\sigma}\frac{\partial}{\partial p}\right)$$

and $K^2 = k^2 + l^2$, and Equation (3.3) becomes

$$\bar{u}\frac{d\hat{\psi}}{dp} = \frac{d\bar{u}}{dp}\hat{\psi}. \tag{3.5}$$

Equations (3.4) and (3.5) form an eigenvalue problem with eigenvalue K^2, and the eigenvector associated with the most positive eigenvalue corresponds to the external mode. Solutions to Equations (3.4) and (3.5) are computed numerically using centered finite differences in pressure. We checked that a single positive mode ($K > 0$) was always found, although setting the lid at a higher level (pressure below 50 hPa) sometimes yields additional spurious modes as discussed in Held et al. (1985). We normalize the external mode by its value at 300 hPa, which is usually close to the maximum.

The external mode is conventionally computed at a given latitude from zonal- and time-mean wind and temperature data (Held et al. 1985). Here, however, we are particularly interested in the variability of the external mode with longitude, and as noted in the introduction to Held et al. (1985) it is possible to define a zonally localized external mode if the time-mean flow is sufficiently slowly varying in longitude. Therefore, we calculate the December–February external mode along the 40°N latitude circle, for ERA-Interim data and each CMIP5/AMIP model in all scenarios, by averaging atmospheric data between 35° and 45°N and then taking a running mean over 60° longitude intervals. We choose a 60° width as a compromise between a width large enough to correctly represent the mean state across the stationary wavelength that dominates the climate change response (waves with nondimensional zonal wavenumber 5 account for most of the projected 300-hPa meridional wind change; (Simpson et al. 2016) and a width small enough to capture zonal variability in the mean state. Subsurface values are discarded based on each model's mean surface pressure; for each 60°

interval, the surface level used in Equation (3.4) is chosen as the highest pressure level with at least 50% of valid data. We refer to the zonally varying external mode as $\hat{\psi}(p, \lambda)$, where λ is longitude. For comparison, we also calculate the external mode using zonally averaged fields and refer to it as $\hat{\psi}_0(p)$.

The external mode allows us to estimate the near-surface component of the stationary wave response using only as input the flow response in the upper troposphere. We compute estimates of the 850-hPa meridional wind change at 40° (35°–45°N average) in the RCP8.5 (for CMIP) and "amipFuture" (for AMIP) scenarios by multiplying, at each longitude, the mean 300-hPa meridional wind by the external mode component at 850 hPa (recalling that the external mode is normalized to be 1 at 300 hPa) and taking the difference between the two scenarios. So, for CMIP models, this corresponds to

$$\Delta v^{850}_{est}(\lambda) = \hat{\psi}^{850}_{\mathrm{RCP8.5}}(\lambda) v^{300}_{\mathrm{RCP8.5}}(\lambda) - \hat{\psi}^{850}_{\mathrm{hist}}(\lambda) v^{300}_{\mathrm{hist}}(\lambda), \qquad (3.6)$$

where historical values are averaged over the 1976–2005 period and RCP8.5 values over the 2071–2100 period. Using only the historical external mode leads to local differences of less than 5%, and therefore differences in external-mode amplitudes between scenarios is of minor importance in our problem.

Results

Zonal variability of the external mode is shown on Figure 3.3a. Consistent with the findings of Blackmon et al. (1979), the structure of the external mode exhibits a substantial zonal variability. Eastern ocean basins and western continental margins are characterized by much stronger vertical coherence, with larger amplitudes in the lower troposphere, than continents. In particular, the external-mode amplitude at 850 hPa is maximal over the western Mediterranean (≅0.43 when averaged over 10°W–0° in ERA-Interim). Over western North America

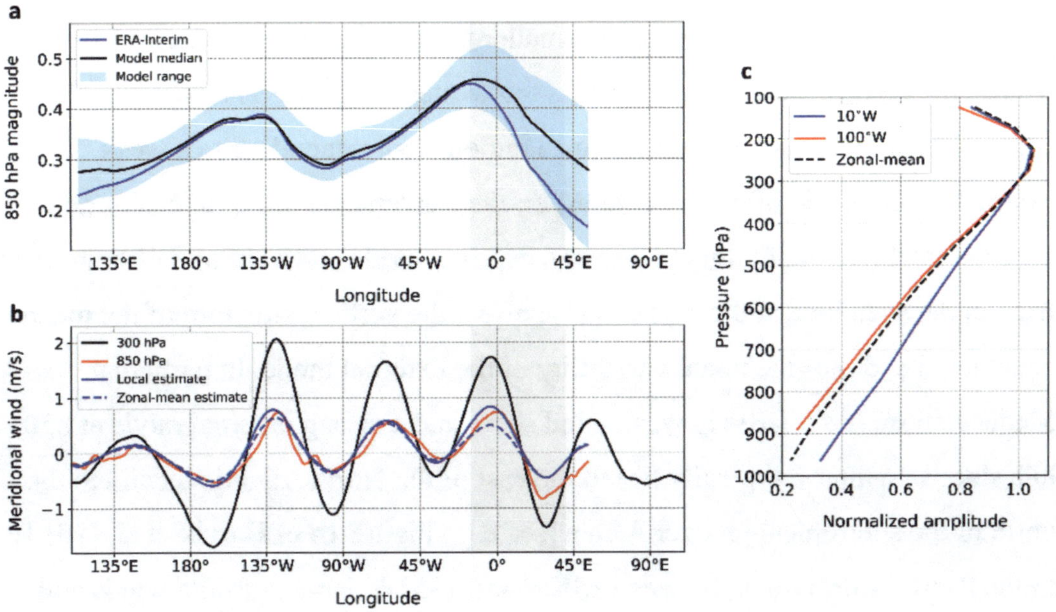

Figure 3.3: (a) Value of the DJF external mode at 850 hPa and 408N as a function of longitude: CMIP5 median in the historical (1976–2005) scenario (black), 95% range (light-blue shading), and ERA-Interim (blue; calculated with 1979–2018 data). The external mode is normalized so that it is equal to 1 at 300 hPa. (b) CMIP5 multimodel mean DJF meridional wind change (35°–45°N average) at 300 (black) and 850 (red) hPa, the estimate based on the external-mode structure (solid blue), and the estimate based on the zonal-mean external mode (dashed blue). In (a) and (b), the Mediterranean region (15°W–40°E) is highlighted in light gray. (c) Examples of external-mode calculations in ERA-Interim at 10°W (blue), at 100°W (red), and with zonal-mean data (dashed black).

(125°–115°W), that amplitude is only 0.32. The ERA-Interim reference external mode is calculated over a time period (1979–2018) that is different from that of CMIP5 models (1976–2005), but results are the same when considering a shorter time period (e.g., 1979–2005) for ERA-Interim data.

The variations with longitude in Figure 3.3a essentially reflect the variability in the vertical structure of the zonal wind profile. Indeed, estimates of the vertical mode based on zonal-mean static stability values show little difference with the Figure 3.3a values. The external mode tends to have similar vertical structure as the mean zonal wind (Held et al. 2002) (Figure 3.3c), and thus the external mode is more barotropic the smaller the vertical shear of the mean zonal wind. This vertical shear is strongest on the western side of ocean basins and adjacent land in the storm track entrance regions (corresponding to a weak external mode at 850 hPa) and weakest on the eastern side of ocean basins and adjacent land (corresponding to a strong external mode at 850 hPa). Differences in friction between land and ocean can also affect the vertical structure of the mean zonal wind and thus the zonal variability of the external mode. In particular, the Mediterranean has relatively weak wind shear and a strong external mode at 850 hPa since in winter it typically lies to the east of the North Atlantic jet and to the north of the subtropical jet over Africa [see e.g., Figure 1b of Harnik et al. [2014]. In the Pacific storm track, between 135°E and 135°W, large 300-hPa winds and weak near-surface winds generate strong shear, resulting in a smaller 850-hPa external-mode magnitude relative to the North Atlantic.

It is interesting to notice from Figure 3.3a, however, that although the external-mode amplitude is generally well-reproduced in models, across the 40°N latitude circle, the single area of significant discrepancy is the central and eastern Mediterranean, up to the beginning of the Central Asian mountain chains (10°–55°E). There, models almost systematically overestimate the low-level external-mode amplitude, on average by 40%. This suggests that, even if the upper-level circulation change projected by models is realized, the regional surface circulation response will be weaker than in model projections. In addition, intermodel scatter in external-mode magnitude in this region is much larger than elsewhere, which may point to differences in the resolution of orography and in the effect of orography on the large-scale flow. The overestimation of the external-mode component over the eastern Mediterranean seems related to a large underestimation of zonal wind shear by GCMs over northern Africa and the Middle East, along the northern border of the North African jet (Figure 3.4). Weaker shear is consistent with an overestimation of external-mode magnitude at low levels, although biases in mean 300-hPa zonal wind may also matter (see Equation 3.4). In addition, Figure 3.4 also shows that CMIP5 models tend to overestimate vertical shear in the North Atlantic midlatitude jet over northwestern Europe, consistent with their average bias toward a too zonal storm track during winter (Simpson et al. 2020). The bias of too little shear further south may be related to a bias of the jet being too rarely in its southward position (Simpson et al. 2020), though a comprehensive analysis of circulation biases over Europe would be required.

This zonal variability has important implications for the low- level circulation response. Figure 3.3b shows the two estimates of the 850-hPa meridional wind change, using either the zonally varying or the constant external mode, averaged across all models. In general, the two estimates are very consistent with the 300-hPa model response. The magnitude of the 850-hPa change is correctly captured, even with the zonal-mean external mode, except where the external-mode amplitude reaches a local maximum, i.e., over the western Mediterranean (~10°W) and off the coast of Western North America (~135°W). Over each of these regions, the peak in 850-hPa external mode amplitude coincides with a peak in 300-hPa meridional wind change, so that important discrepancies are found between the zonal- and local-based meridional wind estimates. In the western Mediterranean in particular, it is necessary to use the local external-mode value to account for the full magnitude of the 850-hPa

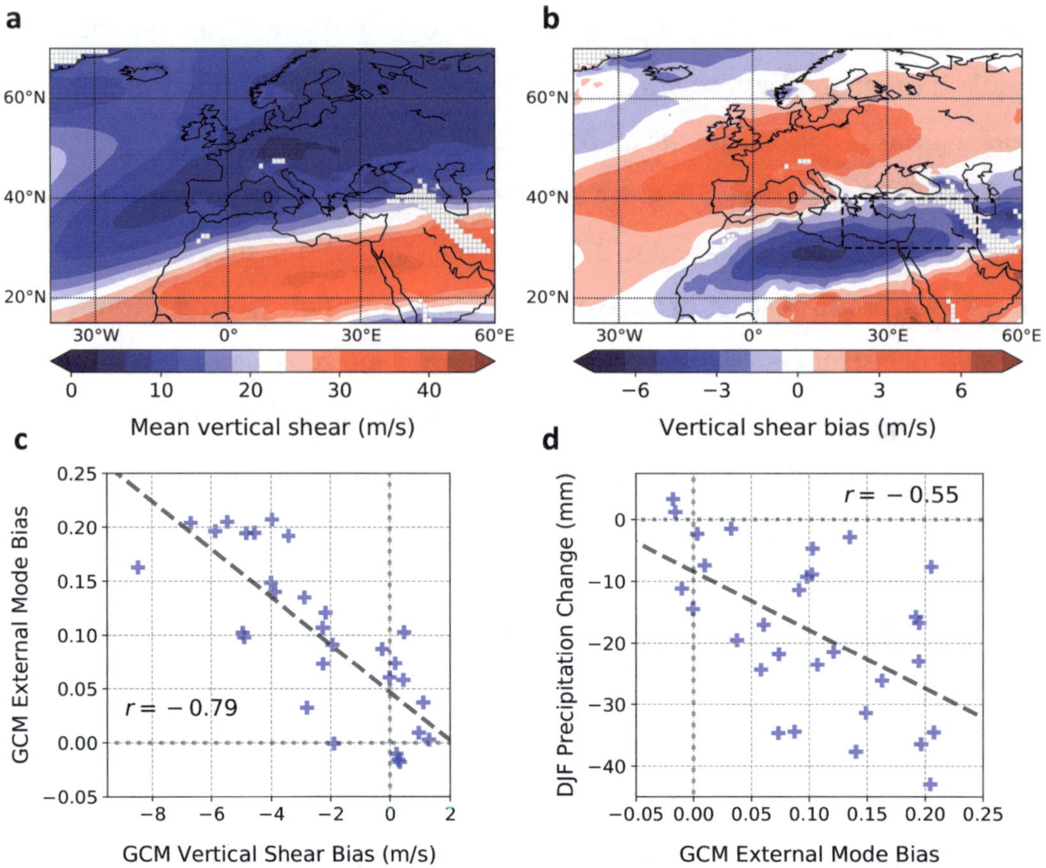

Figure 3.4: (a) DJF Mediterranean mean vertical wind shear, from ERA-Interim. (b) CMIP5 multimodel-mean bias in DJF Mediterranean vertical wind shear. (c) Bias in vertical wind shear over the southeastern Mediterranean region [30°–40°N, 20°–50°E; see dash-outlined rectangle in (b)] against bias in the eastern Mediterranean (16°–46°E) external-mode magnitude at 40°N across CMIP5 models. The region for wind shear calculation is selected empirically on the basis of its high correlation with the eastern Mediterranean external-mode bias. (d) Bias in 850-hPa external-mode amplitude over the southeastern Mediterranean region [as in (c)] against Mediterranean (30°–45°N, 10°W–40°E) precipitation change under RCP8.5. In (a)–(c), vertical wind shear is defined as 300-hPa zonal wind minus 850-hPa zonal wind. In (c) and (d), best-fit linear regression lines are shown in dashed black.

wind change.

It is important to point out that the estimate based on Equation 3.6 does not explicitly take the potential vertical tilt of the stationary circulation into account. A small phase shift can indeed be detected on Figure 3.3b between the 850-hPa response and its external-mode-based estimate, for instance between 90° and 30°W, that could be related to vertical tilt in the time-mean circulation. However, the magnitude of the 850-hPa response is still correctly estimated. Our argument is therefore still valid, even if for an accurate estimate of the low-level response at a given longitude, one should also take into account any tilts in the circulation.

It must also be stressed that the local external mode only gives information about the local vertical structure of the time-mean circulation. Therefore, in the climate change context, this approach can only give information about zonal variability in the vertical structure of the circulation response, not zonal variability in the circulation response itself. In particular, the circulation estimate from Equation 3.6 relies on taking the 300-hPa response in CMIP5 models as given. Future work should focus on providing a theoretical understanding, for instance with a stationary wave model in an idealized setting, on the extent to which zonal variations in the local external mode can lead to variations in the amplitude of the stationary wave in the upper troposphere (e.g., at 300 hPa) in addition to variations near the surface.

In addition to the multimodel mean, intermodel spread is also well explained by this approach. Figure 5 shows the values of the local external-mode-based estimate of 850-hPa wind change against those given by models. We consider both the average and the maximum change. There is excellent agreement

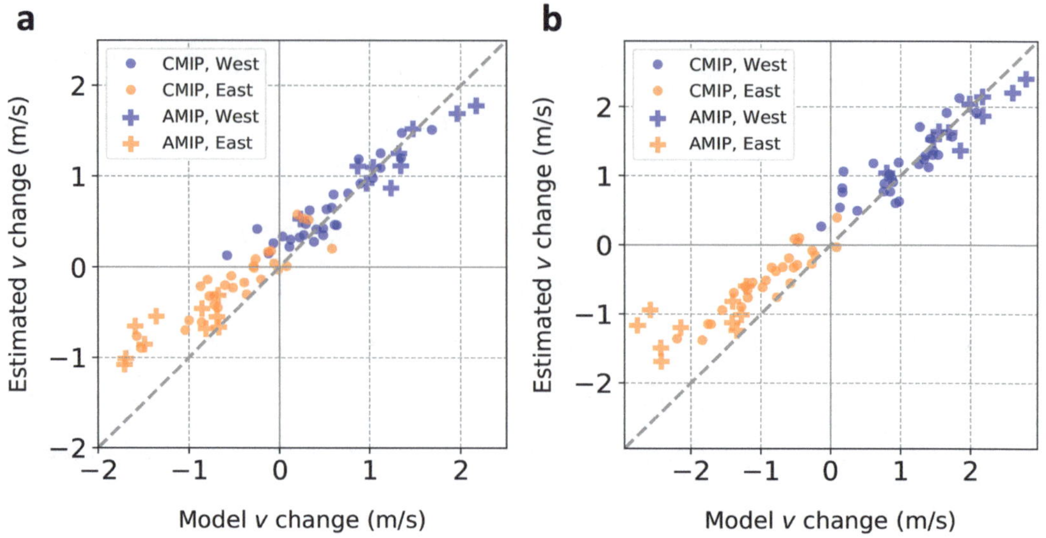

Figure 3.5: The 850-hPa meridional wind change (35°–45°N average) in the western (15°W–15°E; blue) and eastern (16°–46°E; orange) Mediterranean in CMIP5 (circles) and AMIP (crosses) models. The changes estimated through the external mode [Equation 3.6] are plotted vs the changes in the models. Shown are (a) changes averaged over each region and (b) the maximum changes in each region.

in both cases, although for models falling at the extremes the external-mode-based estimate tends to underestimate the wind response, particularly in the eastern Mediterranean. The larger changes found in the AMIP simulations are also well accounted for by this method, i.e., they mainly result from larger circulation changes in the upper troposphere in AMIP simulations. Differences in external-mode magnitude across models are found to play a smaller role in the spread of 850-hPa meridional wind projections when compared with the scatter in upper-tropospheric trends.

Since precipitation projections for the Mediterranean are strongly related to the strength of low-level circulation changes (Zappa and Shepherd 2017, Chapter 2), the external-mode approach can also account for intermodel spread in precipitation trends. Additionally, for a given 300-hPa response, a larger external-mode amplitude over the Mediterranean will trans- late into larger low-level circulation changes. It is therefore interesting to test whether CMIP5 GCM biases in external mode amplitude and vertical wind shear are linked to the magnitude of precipitation projections in the Mediterranean. Figure 4d shows that it is indeed the case: larger external mode biases over the eastern Mediterranean tend to be associated with more severe declines in winter precipitation. There is naturally a substantial spread, given that the strength of the 300-hPa response is not fixed for a given shear bias. Still, the relationship is significant (p value < 0.05) and relies on a clear physical mechanism. Therefore, biases in shear across climate models can be seen as an emergent constraint (Hall et al. 2019) to reduce the spread in future Mediterranean precipitation trends during winter.

Although the GCMs tend to overestimate the barotropicity of the circulation over the eastern Mediterranean, it is also clear from Figure 3.3b that in that region, both external-mode-based meridional wind estimates account for only half or so of the observed wind change in models (the difference is, on average, 0.3 m s^{-1}). To test whether this conclusion is sensitive to our choice of central latitude, we repeat the external-mode analysis over 5° latitude bands centered at 37.5° and 42.5°N. Results in Figures 3.6a and 3.6b show that while the external-mode approach ac- counts for much of the 850-hPa meridional wind response over 40°– 45°N, it largely underestimates it between 35° and 40°N, explaining the original bias identified in Figure 3.3b over the full 35°–45°N interval. For further understanding, we consider the vertical structure in the zonally averaged meridional wind change over the eastern Mediterranean (Figure 3.6c). Meridional wind trends are largest in magnitude in the upper troposphere, near 200 hPa, where they show a pronounced decline around 45°N consistent with

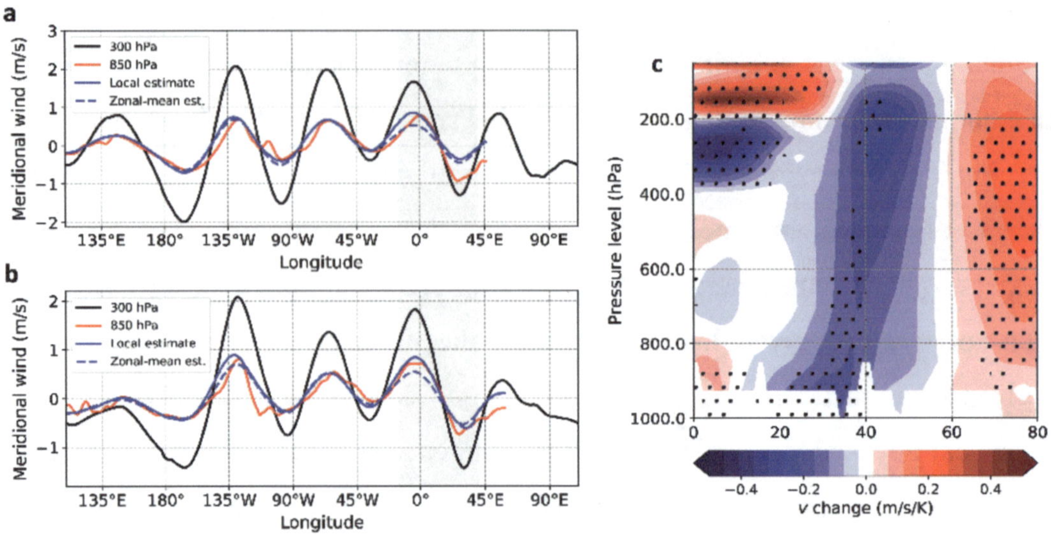

Figure 3.6: CMIP5 multimodel mean DJF meridional wind change averaged over (a) 35°–40°N and (b) 40°–45°N, at 300 (black) and 850 (red) hPa, the estimate based on the external-mode structure (solid blue), and the estimate based on the zonal-mean external mode (dashed blue). In (a) and (b), the Mediterranean region (15°W–40°E) is highlighted in light gray. (c) CMIP5 multimodel mean change in meridional wind averaged over 16°–46°E under RCP8.5 (2071–2100 minus 1976–2005), normalized by each model's global-mean temperature change. Stippling indicates that 80% or more of models agree on the sign of the change.

Figure 3.2b. The amplitude of the trends decreases with pressure, but not all the way to the surface. Rather, a robust secondary local maximum in meridional wind change can be found at the surface, near 35°N. This points to the existence of an additional forcing at the surface, discussed in the next section, that enhances the circulation response linked to upper-tropospheric trends and that can explain the discrepancy between the model and estimated wind responses in Figures 3.3 and 3.6a.

Regional circulation response to surface forcing

In addition to upper-tropospheric circulation trends and their impact over the Mediterranean, the regional circulation also responds to the weaker rate of warming of Mediterranean SSTs compared to land, a robust feature over oceans under global warming (Sutton et al. 2007; Byrne and O'Gorman 2018). Chapter 2 argued, based on regional simulations, that in the Mediterranean in winter, the

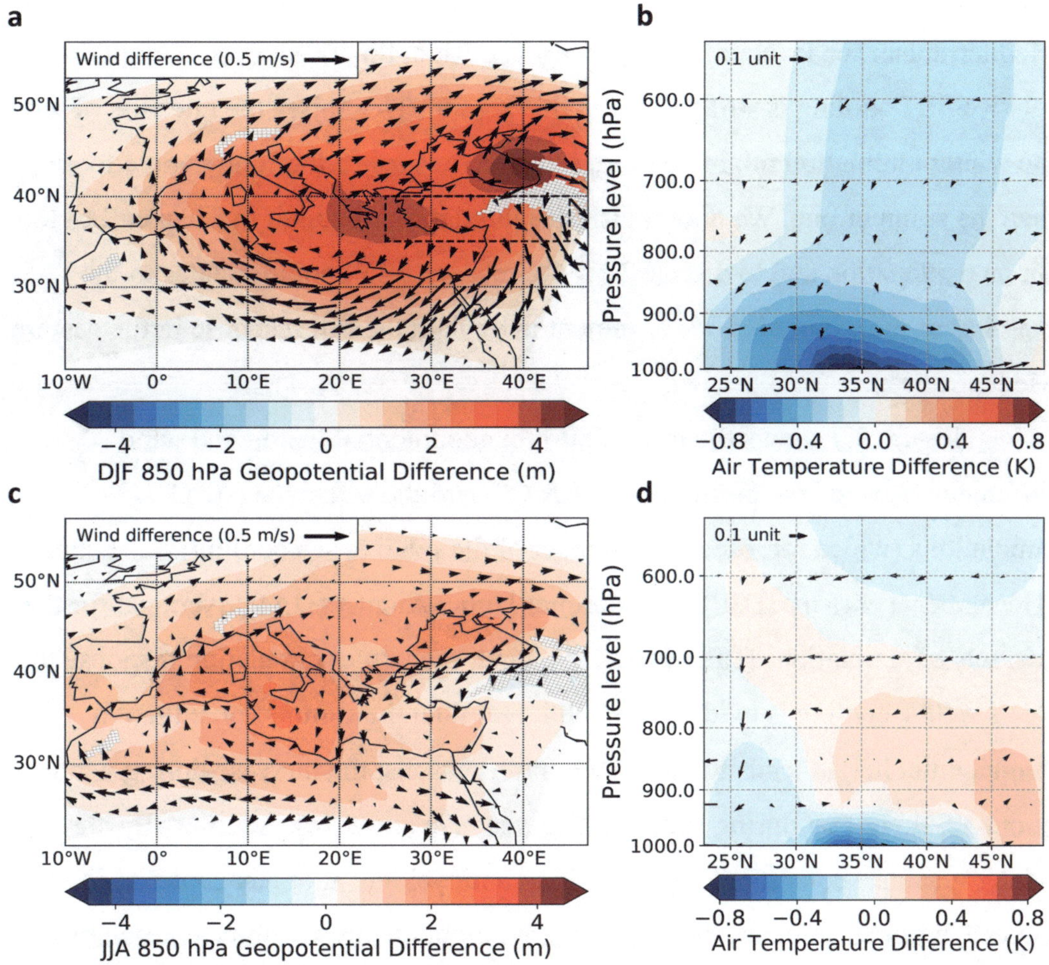

Figure 3.7: Difference in (a),(c) 850-hPa geopotential height and wind and (b,(d) zonally averaged (0°–40°E) air temperature, meridional wind, and pressure velocity, between the "0C" and "+1.5C" simulations ("0C" minus "+1.5C"; average of ERA and MPI runs) for (top) winter (DJF) and (bottom) summer (JJA). In (b) and (d), circulation vectors are scaled such that 1 horizontal unit = 1 m s^{-1} and 1 vertical unit 5 0.2 hPa h^{-1}.

relative cold SST anomaly due to enhanced land warming triggers a regional circulation response with higher pressure downstream of the anomaly and anomalous northeasterlies in the eastern Mediterranean. However, GCM projections for summer clearly show a robust decrease rather than an increase in Mediterranean sea level pressure (Giorgi and Lionello 2008), even though the relative SST cooling is still present in this season. This raises the question of why the winter circulation might be more sensitive to surface thermodynamic forcing than the summer one. We now revisit the simulations detailed in Chapter 2 to look for an explanation and to discuss how the relative SST cooling relates to the inability of the external-mode argument to explain the full response in the eastern Mediterranean.

Figure 3.7 shows the mean 850-hPa geopotential height and wind difference between the perturbed ("+1.5°C") and the reference ("0°C") simulations (which represents the impact of the relative sea cooling) for winter [December–February (DJF)] and summer [June–August (JJA)], averaged for the two sets of simulations (ERA and MPI). First, the winter circulation response to relative SST cooling is evidently much larger than the summer one: though geopotential height tends to increase in both seasons, the increase during winter is about 3.5 times that during summer (Figures 3.7a and 3.7c). Additionally, the winter response consists of a well-defined anticyclonic anomaly centered south of Greece, while in summer, the geopotential height change exhibits a more uniform increase over the Mediterranean. The associated summer circulation response is small everywhere ($\leq$ 0.1m s^{-1}).

For further analysis, we look at the temperature and circulation response on pressure levels (Figures 3.7b,d). The striking difference is that during winter,

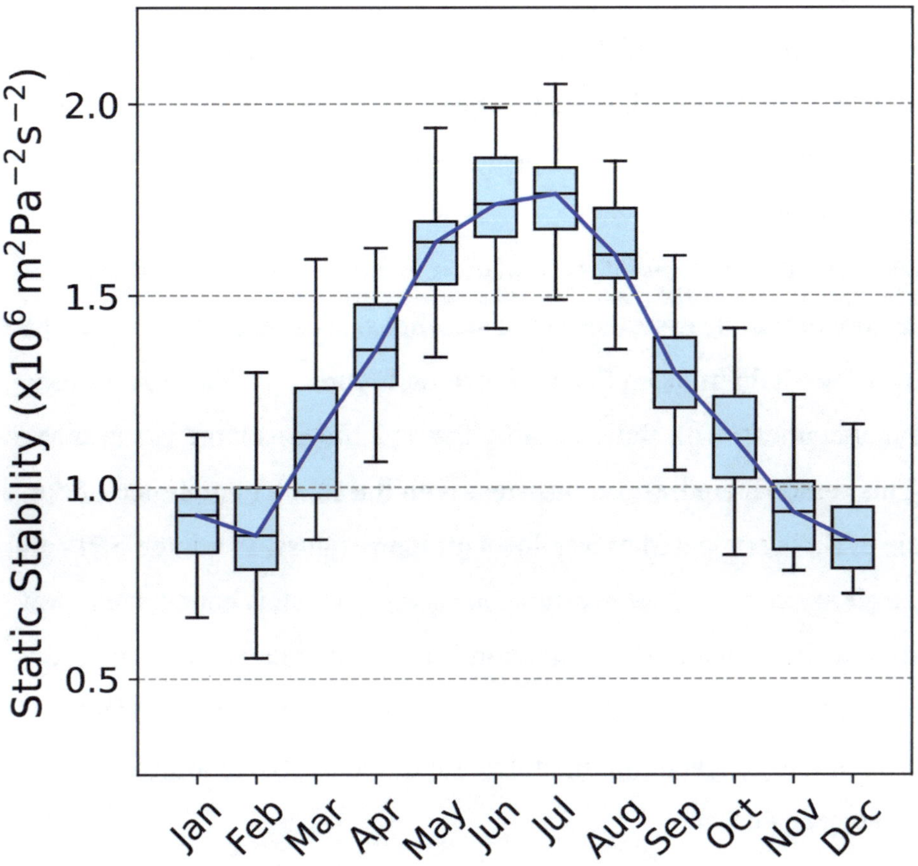

Figure 3.8: Annual cycle of 925-hPa Mediterranean-average (30°–45°N, 10°W–45°E) static stability, from ERA-Interim (1979–2018). Boxes represent the 50% range over the 40 years of data, and whiskers show the full range.

the relative cooling extends much higher into the lower troposphere (up to 850 hPa, and even weakly above that) than in summer, when it is confined to the very lowest model level. As expected, there is little to no circulation response below 800 hPa in summer, whereas at these levels in winter there is noticeable divergence and weak descent.

To determine why the winter circulation is more sensitive to relative SST cooling, we now turn to near-surface static stability s. Its annual cycle at 925 hPa averaged over the Mediterranean Sea is shown on Figure 3.8a. There is a strong winter/summer contrast, with static stability that is 2 times as large in summer as in winter. This seasonal contrast is consistent with the smaller amplitude of the annual cycle of SSTs compared to low-level air temperature. In winter, SSTs and near-surface air are warmer than overlying air layers, which is conducive to low static stability, while in summer the situation is reversed: relatively warmer low-level air temperatures are associated with enhanced static stability. The large summer static stability prevents communication of the relative cooling to levels higher than the near surface.

It is important to put this analysis in the context of the regional circulation. The regional circulation is dominated during summer by the large-scale descent forced by the Indian monsoon to the east (Rodwell and Hoskins 1996), which con- tributes to enhanced static stability of the air column. There is no such forced signal during winter: the area of strong descent near the regional Hadley cell edge is over the Sahara Desert, well to the south of the Mediterranean (Figure 3.9). Therefore, the Mediterranean Sea is located enough to the north of the Hadley cell such that the warm SSTs can trigger large-scale ascent over the sea, a striking feature of the vertical circulation (Figure 3.9a). This large-scale ascent in turns

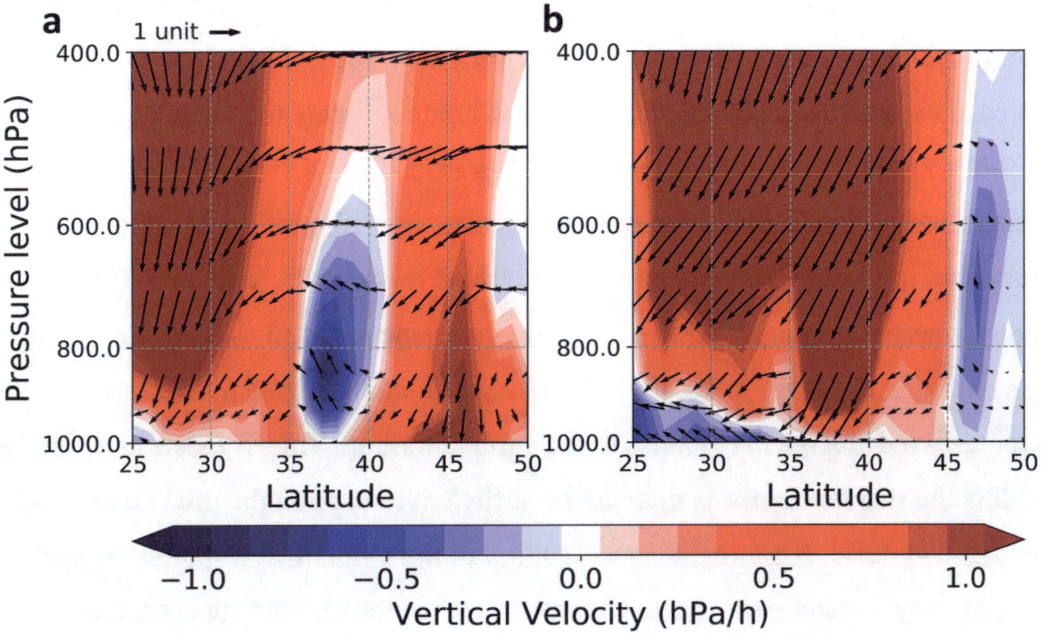

Figure 3.9: Mediterranean (10°W–45°E) zonal-mean (a) DJF and (b) JJA vertical and meridional circulation: vertical velocity (color shades) and wind in the meridional-pressure plane (arrows), in ERA-Interim (1979–2018). Circulation vectors are scaled such that 1 horizontal unit = 3ms^{-1} and 1 vertical unit = 1 hPa h^{-1}.

helps propagate the cooling anomaly upward into the troposphere in our simulations, whereas it remains confined near the surface during summer (Figures 3.7b,d). The combination of the descent and large static stability in summer explains why there is no associated anomalous circulation above the surface since there is no heating anomaly to compensate (Hoskins and Karoly 1981).

In our simulations, the winter relative SST cooling leads to an anomalous high pressure that is centered to the south and east of the anomalous high in the multimodel CMIP5 simulations (cf. Figure 3.1a with Figure 3.7a). There are strong anomalous northerlies to the east of the basin, around 30°–35°N (Figure 3.7a). To the west, the meridional wind response is weaker, consistent with the fact that the stationary wave response explains almost all of the magnitude in meridional wind trends west of 15°E (Figure 3.3b). Thus, the surface forcing helps to explain the inability of the external-mode estimate to account for the full 850-hPa meridional wind response in GCMs around 35°N. The difference in the eastern Mediterranean between projected and estimated meridional wind responses in the 35°–40°N band is concentrated between 25° and 45°E (Figure 3.6b), where it averages 20.5m s^{-1}. Within that same area (35°–40°N, 25°–45°E; see rectangle in Figure 3.7a), the MRCM experiments indicate a 20.23m s^{-1} average wind change in response to the surface forcing. This suggests that simply adding the two mechanisms may not be sufficient to explain the total circulation response at 40°N. A nonlinear superposition of the dynamical responses to each forcing may explain part of that discrepancy (Chapter 2). That superposition is made more complex by the difference in the latitudes of maximum upper-tropospheric (45°N) and surface forcing (30°N), which leads to a noticeable vertical tilt in the meridional wind change (Figure 3.6c). Additionally, the use of a different modeling framework, the finer representation of topography, the influence of friction and high topography on the mean circulation (Simpson et al. 2015) all possibly contribute to shaping the low-level circulation response. Further investigation is warranted, particularly since GCMs probably overestimate the dynamical response to upper-level forcing in the first place, because of their bias in external-mode magnitude over the eastern Mediterranean (Figures 3.3a and 3.4c,d).

Conclusions

We set out to explain the dynamics of two important aspects of Mediterranean circulation trends in a warming climate: why the region experiences in winter a larger surface stationary wave response compared to other regions at a similar latitude, and why the response to relative SST cooling is much larger in winter than summer. First, the Mediterranean during winter is characterized by relatively weak vertical shear in the mean zonal wind, being situated to the east of the North Atlantic jet and to the north of the subtropical jet over Africa. This weak vertical shear imprints on a local external mode (defined over 60° in longitude) and makes the stationary wave response in this region relatively barotropic. As a result, the low-level circulation is particularly sensitive to changes in the stationary wave pattern, leading to a pronounced anticyclonic anomaly. Second, although the relative SST cooling is a year-round feature, its dynamical impact is strongly modulated by the large seasonal variations in low-level static stability and the background circulation. Low static stability during winter allows the heating anomaly to propagate from the surface into the lower troposphere and trigger a circulation response, while in summer, high static stability and intense descent forced by the Indian monsoon keep the heating anomaly confined to the surface.

The physical mechanisms behind wintertime Mediterranean climate projections are robust, but there remains high inter-model uncertainty in the low-level circulation response, and our results show this is mostly linked to differences in the upper-level circulation response. The upper-level circulation response has in turn been related to changes in stationary wave propagation (Simpson et al. 2016), and a deeper understanding of these changes will be required to reduce the uncertainty in Mediterranean projections. In particular, determining what controls the amplitude of the stationary circulation response across GCMs and whether it may be affected by biases in the historical circulation remain open questions. Importantly, we find that the low-level regional circulation response connected to stationary wave shifts may be overestimated in GCMs because their external mode magnitude over the eastern Mediterranean is larger than in ERA-Interim. This bias is related to models underestimating vertical wind shear over the eastern Mediterranean, and it is related through an emergent constraint to the future precipitation response for the whole Mediterranean (Figure 3.4d), suggesting that Mediterranean wintertime circulation change, and consequently precipitation decline, may be overestimated by GCMs.

Acknowledgments. We acknowledge the World Climate Research Programme's Working Group on Coupled Modelling, which is responsible for CMIP, and we thank the climate modeling groups for producing and making available their model output. This work was made possible through funding from the Office Chérifien des Phosphates (OCP) through Université Mohamed VI Polytechnique, Morocco.

A version of this chapter first appeared as
Tuel, A., O'Gorman, P. and Eltahir, E., 2021. Elements of the Dynamical Response to Climate Change over the Mediterranean. Journal of Climate, 34(3), pp.1135-1146. (Original article under AMS copyright)

Chapter 4: Understanding climate change over the southwestern Mediterranean using high-resolution simulations

Alexandre Tuel
Suchul Kang
Elfatih A. B. Eltahir

Among all land regions, the Mediterranean has long been recognized as a major climate change hotspot for the coming century (Giorgi and Lionello 2008; Diffenbaugh and Giorgi 2012). The vast majority of climate models point to a shift towards significantly drier conditions during winter and spring, when southern Mediterranean countries receive the bulk of their annual precipitation. Coupled Model Intercomparison Project phase 5 (CMIP5, Taylor et al. 2012) Global Climate Models (GCMs) project that annual precipitation decreases could locally reach 40% under a business-as-usual scenario (RCP8.5, IPCC 2013) (Zappa et al. 2015) If realized, this would severely affect the habitability and stability of the basin in the next decades (Dubrovsky et al. 2014). These changes

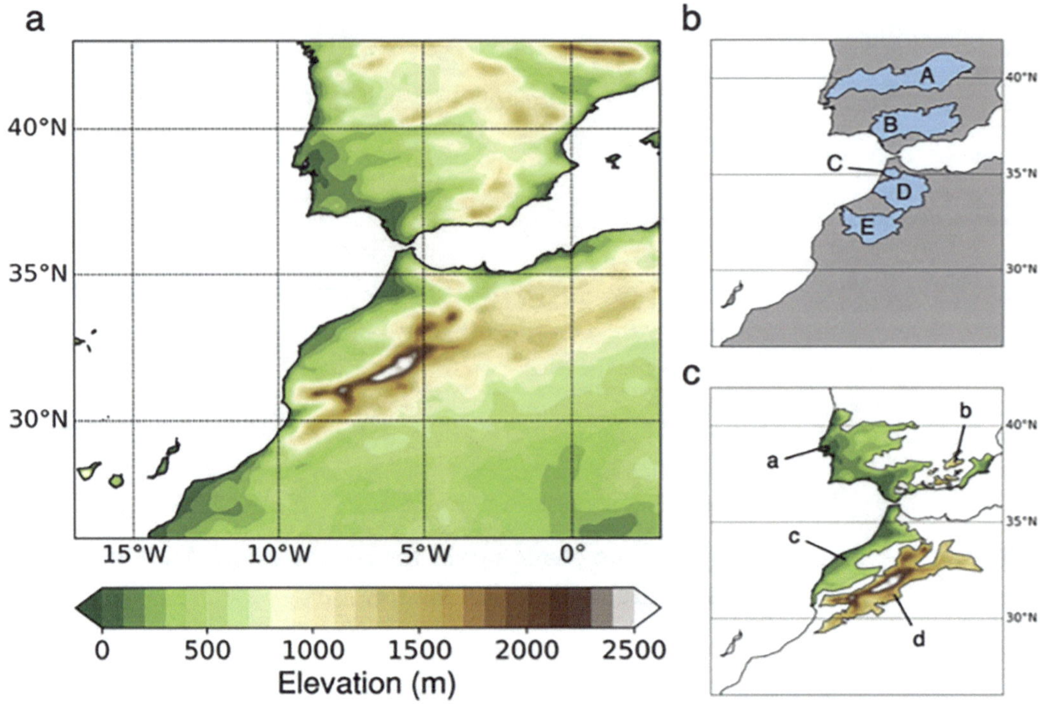

Figure 4.1: a MRCM simulation domain and topography. b Sub-regions used for regional analysis of precipitation projections: A: Tajo, B: Guadalquivir, C: Loukkos, D: Sebou and E: Oum-Er-Rbia watersheds. c Sub-regions used for regional analysis of temperature projections: a: Iberian coastal plain, b: Sierra Nevada, c: Moroccan coastal plain and d: Atlas Range. a, c are defined by altitude < 500 m and b, d by altitude > 1200 m at the MRCM resolution

are bound to have enormous impacts on the regional economy, very dependent on the agricultural sector and water resources. This is particularly true for Morocco and southern Spain (Balaghi et al. 2012), where the decline in precipitation will be highest across the whole Mediterranean (Chapter 2).

However, precipitation over Morocco and Andalusia is strongly variable in space and time, and largely determined by the complex topography, with large maxima over the Rif and Atlas Range peaks, and over Spain's coastal plains and Sierra Nevada (Figures 4.1a, 4.3a). Because of their coarse resolution, GCMs are unable to reproduce these fine-scale, critical details, making the case for the downscaling of climate projections (Gao et al. 2006). Climate data records are very sparse in this area, notably in the mountain regions, and consequently statistical downscaling is hardly an option as it requires long series of observations for calibration. Regional Climate Models (RCMs), on the other hand, are well-suited when reliable records are not available. They can provide more detailed physical insights into the future climate of a region, down to scales that are relevant for water management applications. RCMs have notably been shown to bring significant improvements to the modelling of the complex North African climate (Tramblay et al. 2013; Romera et al. 2015), where the correct simulation of precipitation is challenging due to the high spatio-temporal variability.

Several studies have already evaluated climate change impacts at high resolution over this area. Most used horizontal grid spacings larger than 20 km (Gao et al. 2006; Driouech et al. 2010), with a few studies going down to under 15 km within the framework of the Coordinated Regional Climate Downscaling Experiment database (CORDEX) (Tramblay et al. 2013; Marchane et al. 2017; Fantini et al. 2018; Tramblay et al. 2018). However, these studies usually focus on the Mediterranean as a whole, or on Morocco or the Iberian Peninsula separately. Overall, they have confirmed the trends exhibited by the GCMs, highlighting the region's particular vulnerability to climate change. For instance, Tramblay et al. (2013) predicted a reduction in precipitation of up to 19% and a temperature rise of 1.9 °C by mid-century over the Makhazine catchment in northern Morocco.

Still, past studies have a number of limitations. First, the choice of the forcing GCM is, to the best of our knowledge not discussed, whereas it is well-known that various GCMs have different biases and some are not able to correctly model the climate in given regions (Tebaldi et al. 2005; McSweeney et al. 2015). Bias correction is the preferred option to tackle this problem, but some GCMs may not be able to accurately represent large-scale climate drivers and known teleconnections, thus diminishing our confidence in their projections. Besides, as noted by Tramblay et al. (2013), bias correction for precipitation yields unsatisfactory results even in 12 km-resolution RCMs forced with reanalysis data. Second, and most important, the physical mechanisms behind projected regional climate changes are not discussed. Because the Southwestern Mediterranean exhibits such robust drying trends, it is important to put the RCM results in the context of regional and global trends simulated by the coarse-resolution GCM. An important question is to determine to what extent the physical mechanisms behind the drying at the finer RCM scale are consistent with those identified in GCMs (see Chapter 2), and whether they are better represented at higher resolution.

The goal of this study is to conduct high-resolution (12 km) regional climate simulations over the Southwestern Mediterranean to improve the understanding of climate change processes in this region, with a specific focus on the strong wintertime drying. We apply the MIT Regional Climate Model (MRCM, Im et al. 2014) over a region centered over the Gibraltar Strait and extending from southern Morocco to northern Spain (Figure 4.1a). The ability of MRCM to correctly simulate the current regional climate is demonstrated by using boundary conditions guided by observations (ERA-Interim reanalysis, Dee et al. 2011). We then downscale projections from three GCMs from the CMIP5 archive (Taylor et al. 2012) under the historical (1976–2005), RCP4.5 and RCP8.5 (2071–2100) scenarios (Van Vuuren et al. 2011). The GCMs are selected based on their ability to represent the main large-scale features of wintertime climate in the region. Future projections are then analyzed to understand the physical processes responsible for the robust regional precipitation decline and to connect them to larger-scale climate trends in the Mediterranean.

Area of Study

Our simulation domain, centered at 34°N, 5°W, covers most of Morocco, Portugal and southern Spain (Figure 4.1a); we refer to it hereafter as Northwest Africa and Iberian Peninsula (NWAIP). It has 166 × 161 grid cells with a horizontal resolution of 12 km, sufficient to resolve much of the complex regional topography. Most of the precipitation occurs over Portugal, western Spain and northwestern Morocco, coastal regions which stand under the influence of North Atlantic storm systems (Hurrell et al. 2003, Knippertz et al. 2003). There, the climate is of the Mediterranean type: winters are mild and wet, and summers hot and dry. In Morocco, the Atlas Range acts as a barrier which slows down westerlies, forces orographic precipitation, and cuts off the moisture supply for regions to its east. The Spanish sierras similarly create a rain shadow over southeastern Spain, much drier than the regions to its west. Still, long periods of drought, at any point during the year, and seasonal lack of moisture are common across the domain. Average annual precipitation ranges from 100 mm south of the Atlas, up to about 800 mm over southern Portugal and Morocco's Rif Mountains. The vast majority of it falls during the wet season, from November to May. Storm events during this period are few, short and irregular (Born et al. 2008), occurring as mid-latitude weather systems drifting south-eastward from the North Atlantic. Their frequency increases with latitude, as one gets closer to the storm track center. A large fraction of the winter precipitation falls as snow over the Atlas Range and Iberian sierras, building up a snowpack which slowly melts over spring, guaranteeing a prolonged water supply for the plains below. South and southeast of the Atlas, the Sahara Desert is characterized by a year-round hot and dry climate with less than 100 mm of rain.

Coupled to this spatial variability, rainfall is also highly variable in time, with an inter-annual coefficient of variation ranging from 25% along the Atlantic coast up to 100% in the most arid regions. The main driver of inter-annual winter precipitation variability in the NWAIP is the North Atlantic Oscillation (NAO), an oscillatory pattern of sea-level pressure between the Azores High to the south and the Icelandic Low to the north (Hurrell et al. 2003). High values of the NAO are associated with a northwards-displaced storm track and dry conditions south of 40°N; negative values, on the other hand, lead to wetter-than-average conditions over southern Spain, Portugal and Morocco.

For a detailed characterization of rainfall and temperature changes across this complex region, we divide the land area in our domain into several sub-regions. Relevant to precipitation projections, we select five major watersheds: the Tajo and Guadalquivir, in southern Spain and Portugal, and the Loukkos, Sebou and Oum-Er-Rbia watersheds in Morocco (Figure 4.1b). The last three provide about 70% of river discharge in Morocco (MEMEE 2009) and their water resources are therefore critical for the country. We analyze both winter (DJF) and spring (MAM) seasons, since each accounts for 30–40% of annual rainfall over most of the NWAIP domain. For temperature projections, we focus on coastal and mountainous (above 1200 m) regions in winter (DJF) and summer (JJA). Specifically, we consider the Atlas Range in Morocco and the Sierra Nevada in southern Spain, two areas characterized by an important winter snowpack and therefore highly vulnerable to warming (Figure 4.1c).

Model description and experimental design

The MIT regional climate model (MRCM)

MRCM (Im et al. 2014) is based on the ICTP-Regional Climate Model Version 3 (RegCM3, Pal et al. 2007). MRCM maintains much of the same structure but incorporates several significant enhancements, including coupling of the Integrated Biosphere Simulator (IBIS) land surface scheme (Winter et al. 2009), a new surface albedo assignment (Marcella and Eltahir 2012), modified boundary layer height and boundary layer cloud schemes (Gianotti 2012), an irrigation scheme (Marcella and Eltahir 2014), an improved convective cloud scheme (Gianotti and Eltahir 2014a) and a convective rainfall auto-conversion scheme (Gianotti and Eltahir 2014b). With its incorporation of new physical schemes or modification of original schemes, MRCM has consistently showed comparable or better performance in simulating key climate features across various regions (e.g., North America, West Africa, Southwest Asia, and the Maritime Continent) compared to various regional climate models. In particular, Im et al. (2014) demonstrated from simulations over West Africa that the use of IBIS as the land surface scheme results in better representation of surface energy and water budgets in comparison to RegCM3′s BATS. Additionally, Pal and Eltahir (2016) improved the skill of MRCM in simulating the key surface processes over the desert of southwest Asia, thanks to a new albedo prescription based on NASA/ GEWEX Surface Radiation Budget (SRB) project and emissivity from NASA MODIS surface emissivity data.

In consideration of the good performances from these previous studies, we adopt the same combination of physical parameters and land surface conditions as in Pal and Eltahir (2016) except for the convection scheme, for which we select the Grell scheme with Fritsch-Chappell closure assumption. This choice is based on results of test simulations among all possible schemes (Emanuel or Grell scheme with either Arakawa/Schubert or Fritsch/Chappell cumulus closure scheme) to select the one that simulates observations best. We run the model at a resolution of 12 km. The following physical schemes are employed in our experiment: the IBIS land surface scheme; the sub-grid explicit (SUBEX) moisture scheme (Pal et al. 2000); the radiation package of the National Center for Atmospheric Research (NCAR) Community Climate Model, version 3 (CCM3) (Kiehl et al. 1996); the nonlocal planetary boundary layer scheme (Holtslag et al. 1990); and the ocean–atmosphere flux algorithm proposed by Zeng et al. (1998).

Experiment design

First, we evaluate the model's performance in simulating key processes of the regional climate by performing a dynamical downscaling of the ERA-Interim reanalysis dataset (Dee et al. 2011) with MRCM over the 30-year 1981–2011 period. In this experiment, MRCM is driven by the 1.5° × 1.5° 6-h ERA-Interim data as lateral boundary conditions and the 1° × 1° weekly National Oceanic and Atmospheric Administration (NOAA) optimum interpolation sea surface temperatures (OISST, Reynolds et al. 2002) over oceans. We select the optimal convection scheme in this setting. This simulation is afterwards referred to as ERA/MRCM.

Spatial patterns and regional means of temperature and precipitation are compared to several datasets for robustness, given that no single dataset may perform best over our whole area: first, to TerraClimate (Abatzoglou et al. 2018),

	Boundary conditions	Scenario	Simulation period
Reanalysis downscaling	ERA-Interim + OISST	N/A	1982-2011
Climate change simulations	MPI-ESM-MR	Historical	1976-2005
		RCP4.5	2071-2100
		RCP8.5	2071-2100
	GFDL-ESM2M	Historical	1976-2005
		RCP4.5	2071-2100
		RCP8.5	2071-2100
	IPSL-CM5A-LR	Historical	1976-2005
		RCP4.5	2071-2100
		RCP8.5	2071-2100

Table 4.1: Summary of the ten MRCM simulations

which is based on station and reanalysis data, and is available from 1950 to present at a 4 km resolution. Second, to the Tropical Rainfall Measuring Mission (TRMM, Huffman et al. 2007) 3B43 version 7 precipitation dataset, which is based on remotely-sensed data corrected with station observations at the monthly timescale, available from 1998 to present at a 0.25° resolution. Third, to the

Iberia01 dataset (Herrera et al. 2019), which covers the 1971–2015 period at a 0.1° resolution and is based on a high density of stations covering the Iberian Peninsula. TRMM has been found to perform well over Morocco, where the station network is sparse (Milewski et al. 2015); TerraClimate has been validated in the Mediterranean, but against Global Historical Climatology network stations only (Abatzoglou et al. 2019).

Second, we perform climate change experiments using the same configuration of MRCM as in ERA/MRCM, but with 6-hourly output from three selected GCMs from the CMIP5 data archives ("r1i1p1" ensemble members) used as lateral boundary conditions. The selection process is described in Sect. 3.3. The 1975–2005 period forced by historical GHG concentrations is taken as reference, and compared with the 2070–2100 period forced by the Representative Concentration Pathways (RCP) 4.5 and RCP8.5 emission scenarios (Van Vuuren et al. 2011). In total, nine 31-year GCM runs are dynamically downscaled with MRCM.

In all simulations, the first year is discarded as spinup, leaving 30-year projections (1976–2005 for reference, 2071–2100 for future and 1982–2011 for ERA/MRCM) to be analyzed. We also statistically correct the mean biases of surface temperature and precipitation from MRCM simulations at the monthly time-scale with a parametric quantile mapping technique (Sect. 3.4). A summary of the ten MRCM runs is given in Table 4.1.

GCM selection

We select suitable GCMs to use as forcing boundary conditions for MRCM. GCMs are selected among all CMIP5 participating models for which 6-hourly data is available, according to the following methodology:

1. First, given the dominant role played by the NAO in driving inter-annual precipitation variability over our domain, we first retain thirteen models that are able to capture some of the main features of the NAO, based on the comprehensive analyses by Davini and Cagnazzo (2014) and Wang et al. (2017), and for which 6-h output data is available.
2. Second, we exclude the GCMs that are unable to simulate the impact of the NAO on southwestern Mediterranean precipitation (15°W–0°/30–40°N) during winter. This impact is quantified by the correlation of the DJF NAO index and concurrent precipitation (Figure 4.2). Two procedures are tested to compute the NAO index in GCMs: first, as the time series of the first principal component of DJF sea-level pressure (SLP) over the North Atlantic (90°W–40°E and 20–80°N), and second, as the projection of annual DJF SLP fields onto the «observed» NAO pattern, defined as the first principal component of DJF SLP in the NCEP-NCAR reanalysis (Kalnay et al. 1996) (Figure 4.2a–d). Both approaches yield similar results, though each has its limitations, as discussed in Davini and Cagnazzo (2014). We also assess the ability of GCMs to reproduce the spatial patterns of DJF SLP correlation to Western Mediterranean precipitation, based on spatial correlation coefficients (Figure 4.2e–h).
3. Finally, the historical simulations from the selected GCMs are evaluated over the domain, by looking at climate variables during winter and spring relevant for the present analysis: temperature, relative humidity and precipitation. Spatial patterns and regional means are compared to the datasets discussed in Sect. 3.2.

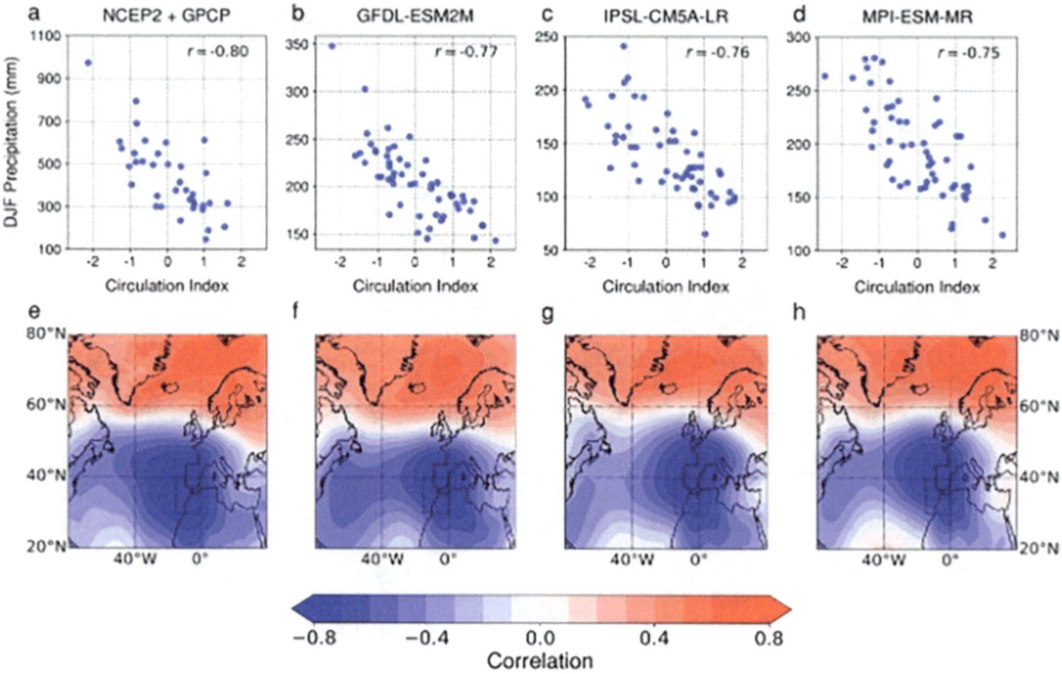

Figure 4.2: a–d DJF Western Mediterranean (15°W–0°/30–40°N) precipitation against concurrent NAO circulation index (see text for definition), with correlation coefficients indicated at top-right; and (e–h) Pearson correlation of DJF SLP to concurrent Western Mediterranean precipitation, in (a, e) GPCP and NCEP-NCAR reanalysis (1979–2017); (b, f) GFDL-ESM2M; (c–g) IPSL-CM5A-LR; and (d, h) MPI-ESM-MR. For (b–h) data covers the 1950–2004 period

We select three GCMs at the end of this screening process: GFDL-ESM2M (Dunne et al. 2012), IPSL-CM5ALR (Dufresne et al. 2013) and MPI-ESM-MR (Giorgetta et al. 2013). These three GCMs generally show the best or close to best performance for the criteria introduced in step 2. Although they all suffer from an evident dry bias (common in fact to most, if not all, CMIP5 GCMs, partly connected to the poor representation of the regional orography, which plays an important role in forcing winter precipitation), they accurately reproduce the SLP-precipitation relationship during winter.

Bias correction

Output from regional climate modelling experiments contains systematic biases arising, among others, from lateral boundary conditions and imperfect representation of physical processes (Liang et al. 2008; Ehret et al. 2012). To reduce this bias, we apply a parametric quantile mapping (QM) method based on cumulative distribution functions (CDF) to monthly-mean temperature and precipitation at each grid point in the domain (Li et al. 2010). Normal (gaussian) and gamma distributions are commonly used for representing the probability density function of temperature and precipitation, respectively. We fit here normal distributions to monthly mean temperatures and gamma distributions to monthly-mean precipitation in both observations and model output, at each grid point and separately for each month. Then, the parametric QM technique consists in correcting the model CDF to match the observed one. As usual in bias correction methods, this transformation is assumed to be remain valid for future climate projections, so the same bias correction is also applied to future temperature and precipitation. For observation datasets, we use for both temperature and precipitation TerraClimate (Abatzoglou et al. 2018), one of the best available consistent long-term and fine-scale gridded observational records. Since Terra-Climate (~ 4 km) and MRCM (12 km) have different grid size, TerraClimate is first regridded to the 12 km MRCM grid resolution.

Results

MRCM performance

We first assess the performance of ERA/MRCM by comparing the simulated precipitation to TRMM, Iberia01 and TerraClimate data, focusing on the climatology over our domain and inter-annual variability for the winter season. While the GCM ensemble mean (ENS/GCM) is able to broadly capture the overall precipitation gradient between the wetter regions over the Iberian

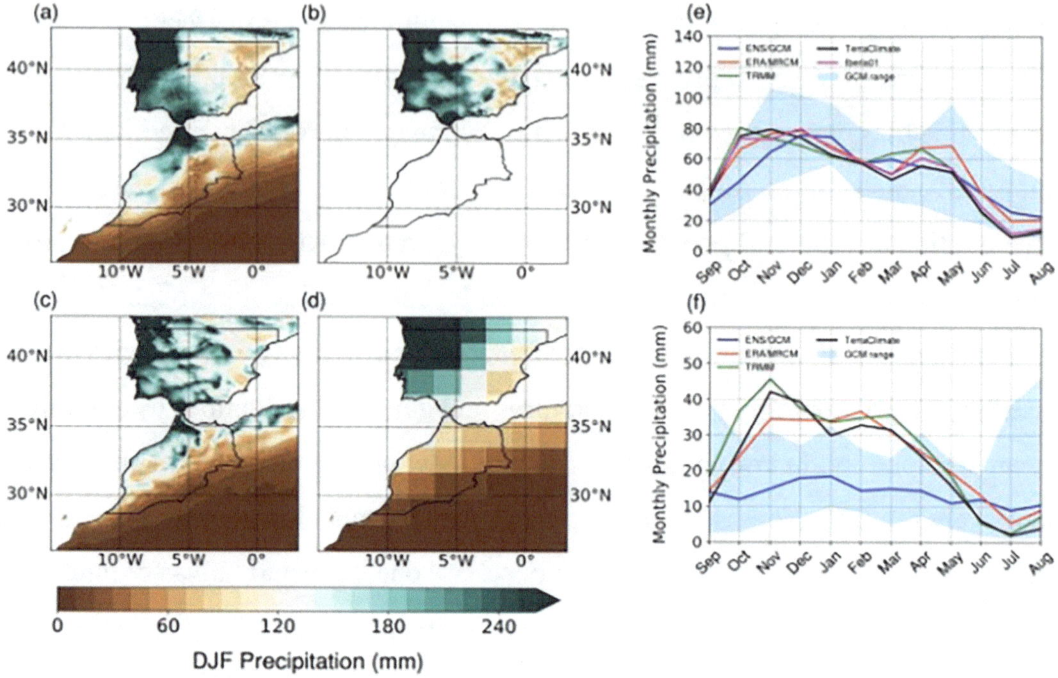

Figure 4.3: a–d Spatial distribution of 30-year climatology of average DJF total precipitation over the NWAIP: a TerraClimate, b Iberia01, c MRCM-downscaled ERA-Interim (ERA/MRCM), d ensemble mean of 13 CMIP5 GCMs (ENS/GCM) (see Table S1). e–f Annual cycles of monthly land precipitation averaged over e the southern Iberian Peninsula and f northwestern Morocco [see black contours in (a–d)], in TerraClimate (black), Iberia01 (magenta, e only), TRMM (green), ERA/MRCM (red) and ENS/GCM (blue). The light blue shading corresponds to inter-GCM spread in the 13-member GCM ensemble

Peninsula and the dry desert over northwest Africa, the small-scale maxima along the African coastlines and over the high mountainous regions are, unsurprisingly, missed (Figure 4.3). Figure 4.3c makes clear the overall dry bias of GCMs over Morocco, expected since high-altitude region. In contrast, ERA/MRCM is in good agreement with the distribution in TerraClimate and Iberia01, both at large- and small-scales. DJF precipitation over the Moroccan and Algerian coastlines is overestimated, as is the case in central Spain, although there TerraClimate exhibits a noticeable dry bias compared to Iberia01, which agrees very well with ERA/MRCM (Figure 4.3b, c). The density of stations used to obtain the data being much higher. South of the Atlas Range, precipitation is also slightly overestimated, possibly a consequence of the incorrect representation of summer small-scale convective systems which bring a large fraction of annual precipitation in this region. Still, MRCM exhibits a pronounced precipitation downscaling skill, especially in regions where orographic forcing dominates. As to annual cycles (Figure 4.3d, e), the ENS/GCM is quite able to capture the seasonal variation of precipitation over the Iberian Peninsula, with its maximum in winter and minimum in summer. The magnitude of wet-season precipitation is reasonably close to TerraClimate, with an evident dry bias all the same. Summer precipitation is quite overestimated, however. By contrast, over northwestern Morocco, GCMs fail to capture the seasonal cycle, showing instead a somewhat constant monthly precipitation rate. The dry bias during winter and spring is evident. Additionally, inter-GCM spread is quite large, especially between July and September when some GCMs produce copious amounts of precipitation, probably the result of exaggerated tropical moisture transports typically responsible for much of the summer precipitation south of the Atlas Range (Born et al. 2008). Compared to ENS/ GCM, ERA/MRCM is in good agreement with TerraClimate, especially over northwestern Morocco, though it fails to fully capture the bimodal distribution with two peaks in December and March as evident in TerraClimate. GCM biases over the Iberian Peninsula, especially a peak in precipitation around May and the underestimation of late fall precipitation, are not fully corrected.

As to inter-annual variability, it is remarkably well captured by ERA/MRCM. Figure 4.4 shows the time series of DJF precipitation anomalies (defined with respect to each series' 1982–2011 mean), over the five selected

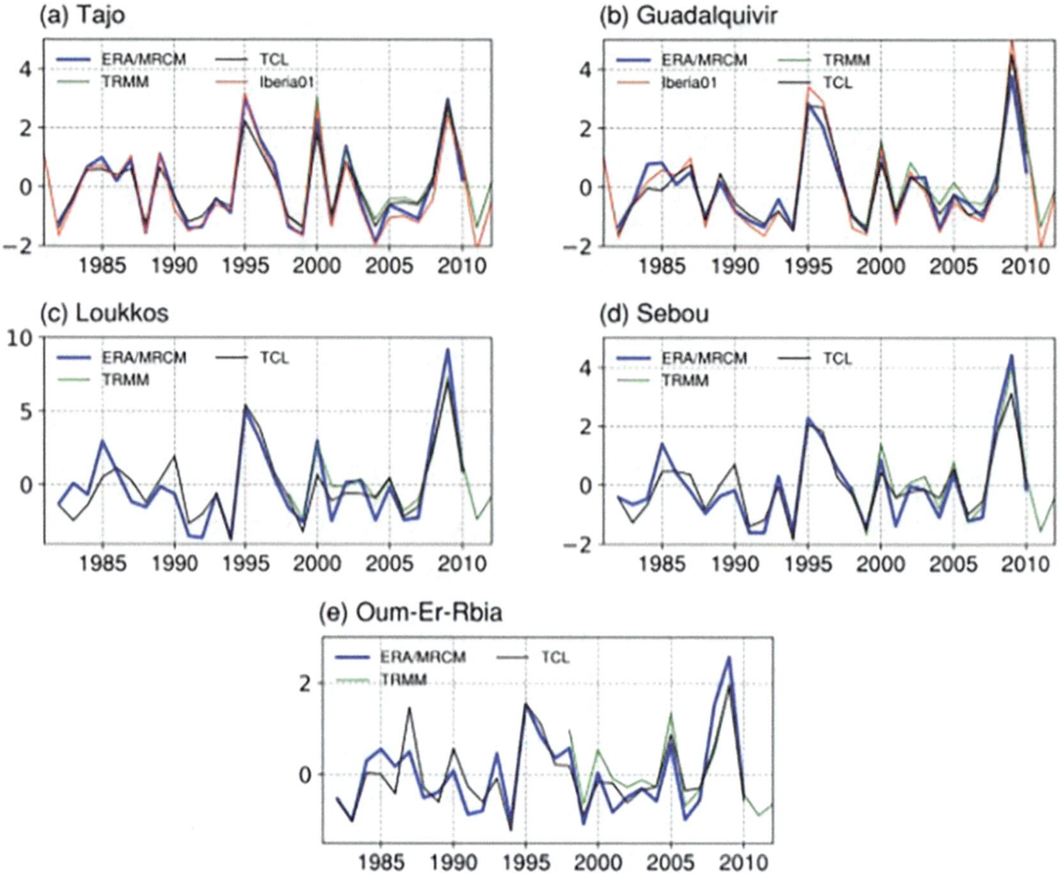

Figure 4.4: Annual DJF precipitation anomalies averaged over the five selected watersheds, in TerraClimate (TCL, black), Iberia01 (red, in a, b), TRMM (green) and ERA/MRCM (blue). Anomalies, in mm/day, are defined by subtracting to each series its 1982–2011 average

watersheds (Figure 4.1b). Anomaly correlation coefficients range from 0.91 to 0.98, which shows that MRCM, for this domain and at the 12 km resolution, is able to correctly simulate the precipitation response to large-scale, NAO-linked storm track shifts present in the boundary conditions.

In summary, ERA/MRCM shows satisfactory performance in simulating both the average and the inter-annual variability of winter precipitation over the NWAIP domain. Even though GCM biases translate to the downscaled simulations, the improved representation of topography helps to compensate in part for GCM's dry bias during the wet season.

Verification of bias corrected MRCM runs

Before analyzing future projections, we validate the performance of the bias corrected GCM-driven MRCM runs (ENS_BC/MRCM) for the reference period, by comparing their results with those from ERA/MRCM and TerraClimate observations. Figures 4.5 and 4.6 show the spatial distribution of the 30-year monthly precipitation and surface mean temperature climatologies for the winter season (DJF) derive from TerraClimate, ERA/MRCM, ensemble mean of three MRCMs (ENS/MRCM) and ensemble mean of three bias-corrected MRCMs (ENS_BC/MRCM).

For precipitation, ENS/MRCM shows good qualitative correspondence with the observed regional distribution (Figure 4.5). However, GCM-driven MRCM runs tend to produce excessive precipitation over higher mountainous regions, particularly over south-central Spain, similarly to the ERA/MRCM run. Wet biases over mountainous areas are indeed a common deficiency in dynamical downscaling, as noted also in other studies (Solman et al. 2008; Gianotti et al. 2012). The exception is IPSL-CM5A-LR, for which the dry bias evident in GCM output (Figure 4.2c) is not fully compensated by the dynamical downscaling. Although it somewhat accurately reproduces long-term NAO dynamics and its relationship to Northern Hemisphere climate variability (Wang et al. 2017), IPSL-CM5A-LR suffers from an "anticyclonic" bias, with a too strong North Atlantic subtropical high that blocks cyclonic activity and leads to the observed dry bias. Regional downscaling cannot compensate for this deficiency.

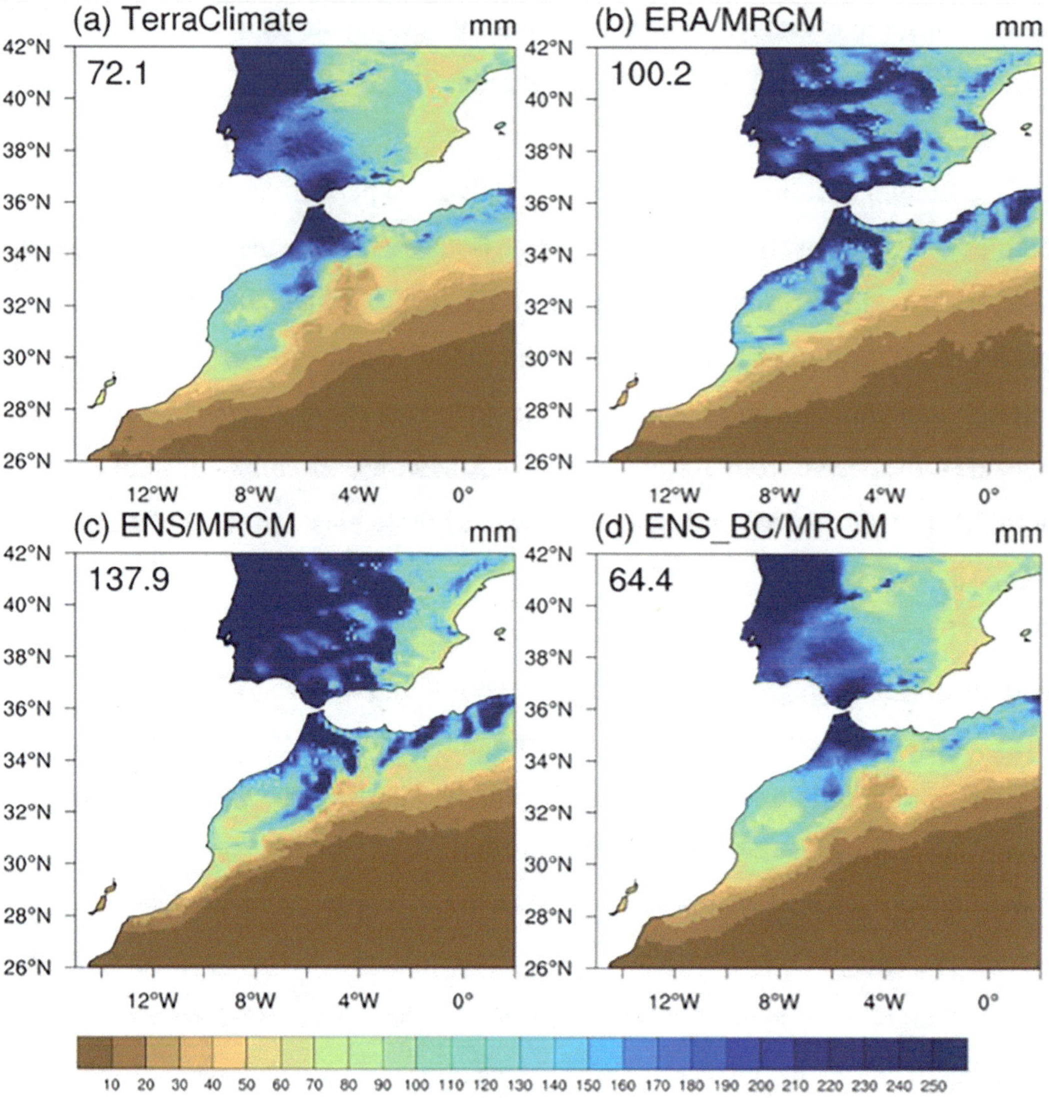

Figure 4.5: Spatial distribution of 30-year climatology of DJF average land precipitation in a: TerraClimate, b: ERA/MRCM, c: ENS/MRCM (ensemble mean of MRCM simulations) and d: ENS_BC/MRCM (ensemble mean of bias-corrected MRCM simulations). Domain-averaged values are indicated at the top left-hand corner of each plot

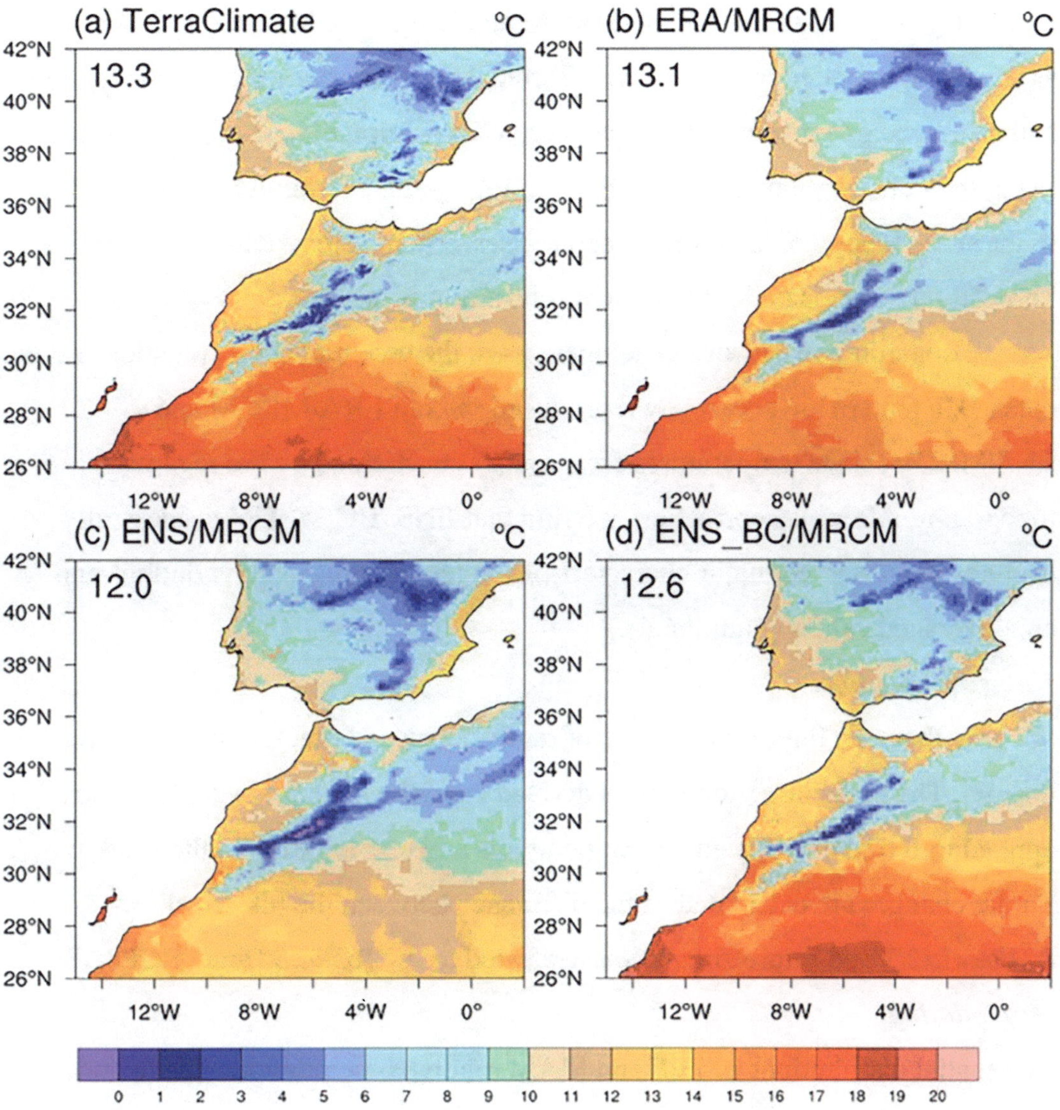

Figure 4.6: Same as Figure 4.5, for surface temperature

By comparison with ERA/MRCM and ENS/MRCM, ENS_BC/MRCM is, as expected, in better agreement with TerraClimate in terms of the spatial pattern and area-averaged precipitation. Bias-corrected precipitation is still somewhat lower than observations over the High- and Anti-Atlas (around 8°W/31°N). Due to the sharp topographic gradients and the scarcity of station data, however, TerraClimate data is undoubtedly less robust in this region.

Regarding mean surface temperatures, the essential characteristics of the mean DJF pattern are accurately reproduced in both ERA/MRCM and ENS/MRCM, since temperatures are strongly constrained by the regional topography. After bias correction, we find that ENS_BC/MRCM successfully captures the observed major characteristics in terms of regional variations and area-averaged values (Figure 4.6).

Future projections

We now focus on future projections of temperature and precipitation over the domain. The bias-correction factors derived from the historical simulation are applied to future monthly-mean temperatures and precipitation in the RCP runs. Future changes are calculated as the difference between the historical ("Historical") and future simulations (referred to as "RCP4.5" and "RCP8.5").

Precipitation

The spatial distributions of DJF and MAM relative precipitation change under RCP4.5 and RCP8.5 are shown on Figure 4.7. A pronounced and robust large-scale precipitation decline over the domain is evident under both scenarios and in both seasons, consistent with the CMIP5 multi-model mean (see Chapter 2) and earlier CMIP5 projections (Giorgi and Lionello 2008). In relative terms, the decline in precipitation is generally larger during spring, at which time the drying

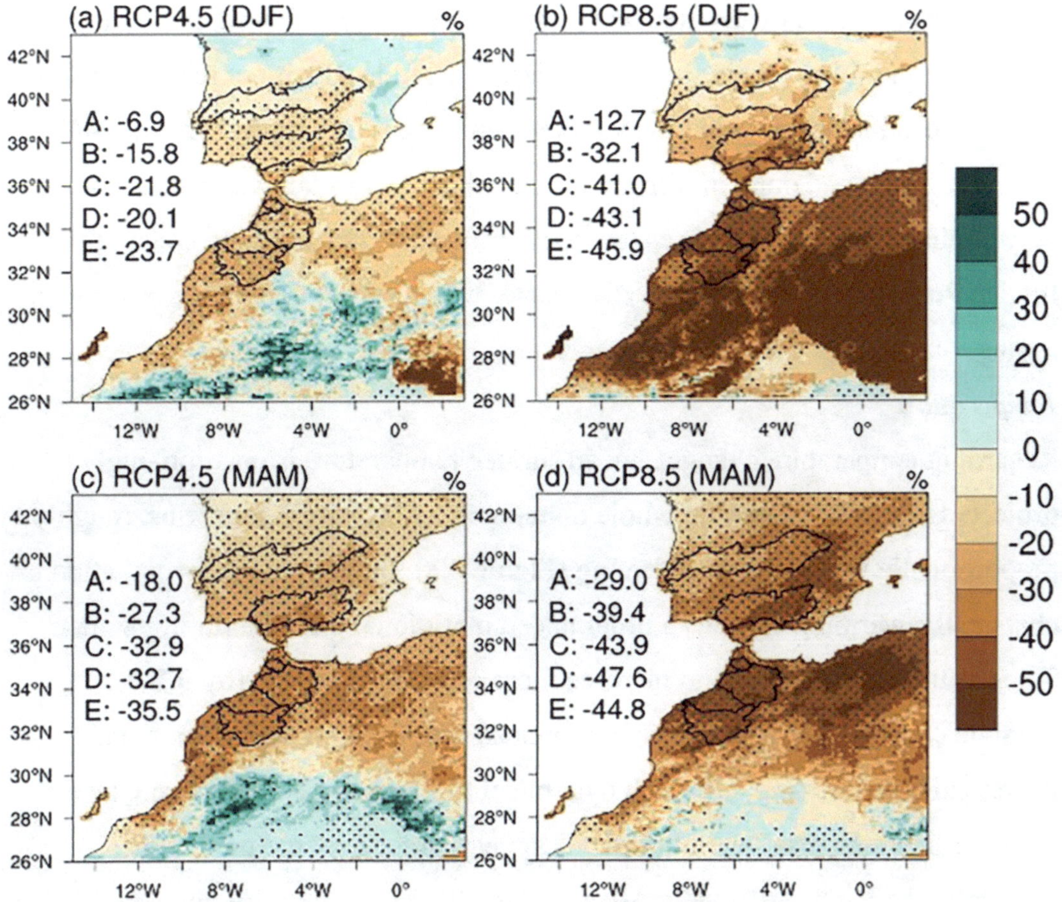

Figure 4.7: MRCM ensemble-mean percentage change in a, b DJF and c, d MAM precipitation, under a, c RCP4.5 and b, d RCP8.5. Mean values averaged over the five watersheds on Figure 4.1b are shown on the left-hand side of each plot. Dots indicate agreement by three MRCM simulations on the sign of the change

is also more extensive, especially over the Iberian Peninsula. While winter precipitation trends in central and Eastern Spain are neither robust nor significant, the situation in spring is quite different. Under RCP8.5, the drying is about twice as large everywhere compared to RCP4.5: from 7–24% to 13–46% (Figure 4.7). In MAM, however, the difference between scenarios is smaller: the decrease in precipitation is very strong even with mitigation (RCP4.5). For the region as a whole, the drying gradually weakens from southeast to northwest. In winter, regions north of 40°N even show slight wetting trends, albeit not robust at all. The African coastline is projected to experience the largest drying by far (− 40/45% under RCP8.5), and quite uniformly too: the three Moroccan watersheds show very similar changes in all cases (seasons/scenarios). The southern coast of the Iberian Peninsula is in an intermediate state, with a robust though weaker winter drying.

Temperature

Regarding temperature changes, mean surface temperature is unsurprisingly projected to increase over the whole domain under both RCP scenarios, roughly proportionally to the radiative forcing (Figure 4.8). Similar to the precipitation change, the warming exhibits a pronounced meridional gradient during winter. Temperatures are projected to increase more over North Africa (by 0.5–1 °C), a possible consequence of its more pronounced precipitation decline through decreased latent heating, although dynamical trends may also play a role (see Sect. 5). In winter also, warming trends appear larger over the mountains, especially the High and Middle Atlas, and the Sierra Nevada (Figure 4.8a, b). This enhanced warming may be connected to the snow-albedo feedback (Rangwala and Miller 2012; Pepin et al. 2015), since these areas all possess extensive snowpacks in the current climate), although this process is not very well resolved at our 12 km scale and deserves a more detailed treatment (Walton et al. 2017). Rather, since mountain regions receive higher precipitation amounts during winter, they will experience larger absolute drops in precipitation, and therefore larger reductions in latent heat flux which will impact on the temperature signal.

In both scenarios, and for all coastal and mountain regions, summer

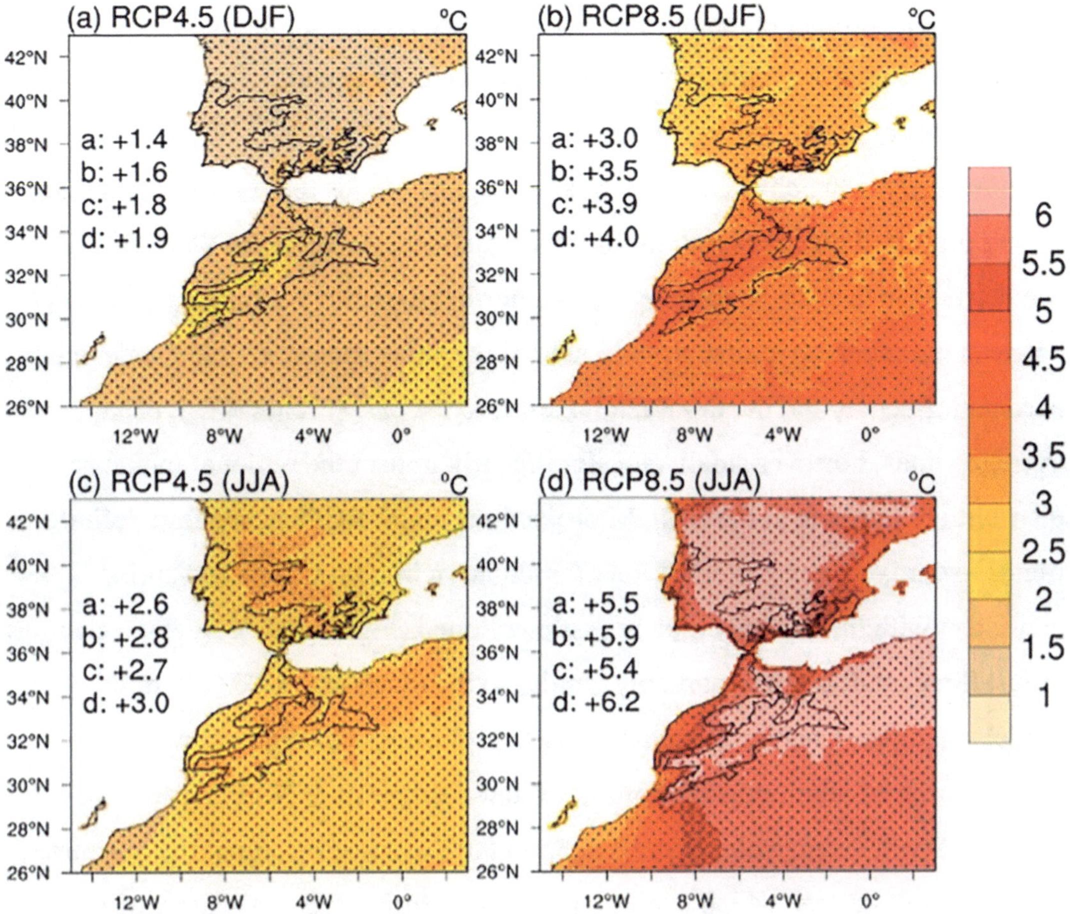

Figure 4.8: MRCM ensemble-mean change in a, b DJF and c, d JJA surface temperature, under a, c RCP4.5 and b, d RCP8.5. Mean values averaged over the four regions on Figure 4.1c are shown on the left-hand side of each plot. Dots indicate agreement by three MRCM simulations on the sign of the change

temperature changes are much higher than in winter (by 1–2 °C). They are characterized by a clear coastline/interior contrast, common to all seasons and scenarios, although here again mountain regions seem to exhibit enhanced warming compared to plains, especially the Atlas and the Pyrenees. As in winter, this may be connected to a reduction in precipitation, here brought by summer convective cells quite common at high elevations in this region (Born et al. 2008).

The reduced warming along the coasts is a classic feature of climate change linked to less pronounced warming of surface air over oceans.

Mechanisms of precipitation change

We now analyze the physical mechanisms responsible for the robust cold-season drying projected for the region in future simulations. Because the Western Mediterranean is characterized by a steep northeast-southwest humidity gradient, between the relatively moist western side of the Atlas Range, which stands under oceanic influence, and the dry Sahara Desert to the east (Figure 4.9a), changes in average winds, however small, can significantly impact the regional moisture distribution. Looking at the MRCM projections under RCP8.5, we find definite trends towards a more easterly kind of circulation in both winter and spring, consistent with the regional picture just mentioned (Figure 4.9b). Average changes in just three models are remarkably similar to the much larger CMIP5 multi-model mean (see Chapter 2), underlining the robustness of regional wind trends under continued anthropogenic forcing. Wind changes over North Africa and southern Spain average about 2 m/s at 850 hPa; this must be compared to average westerlies of 1–5 m/s, which is quite weak, even during the wet season when moisture advection from the North Atlantic is largest (Figure 4.9a). Storms that hit Morocco and the southern parts of the Iberian Peninsula indeed remain quite rare, even under negative NAO conditions.

The robust shift towards enhanced easterlies, which should lead to enhanced advection of dry air, is consistent with regional projections of atmospheric moisture (Figure 4.9b). The northwest African and southern Iberian coastlines stand under relative lows of atmospheric humidity increase, when the rest of the region, particularly northern Spain and Portugal, exhibit much larger

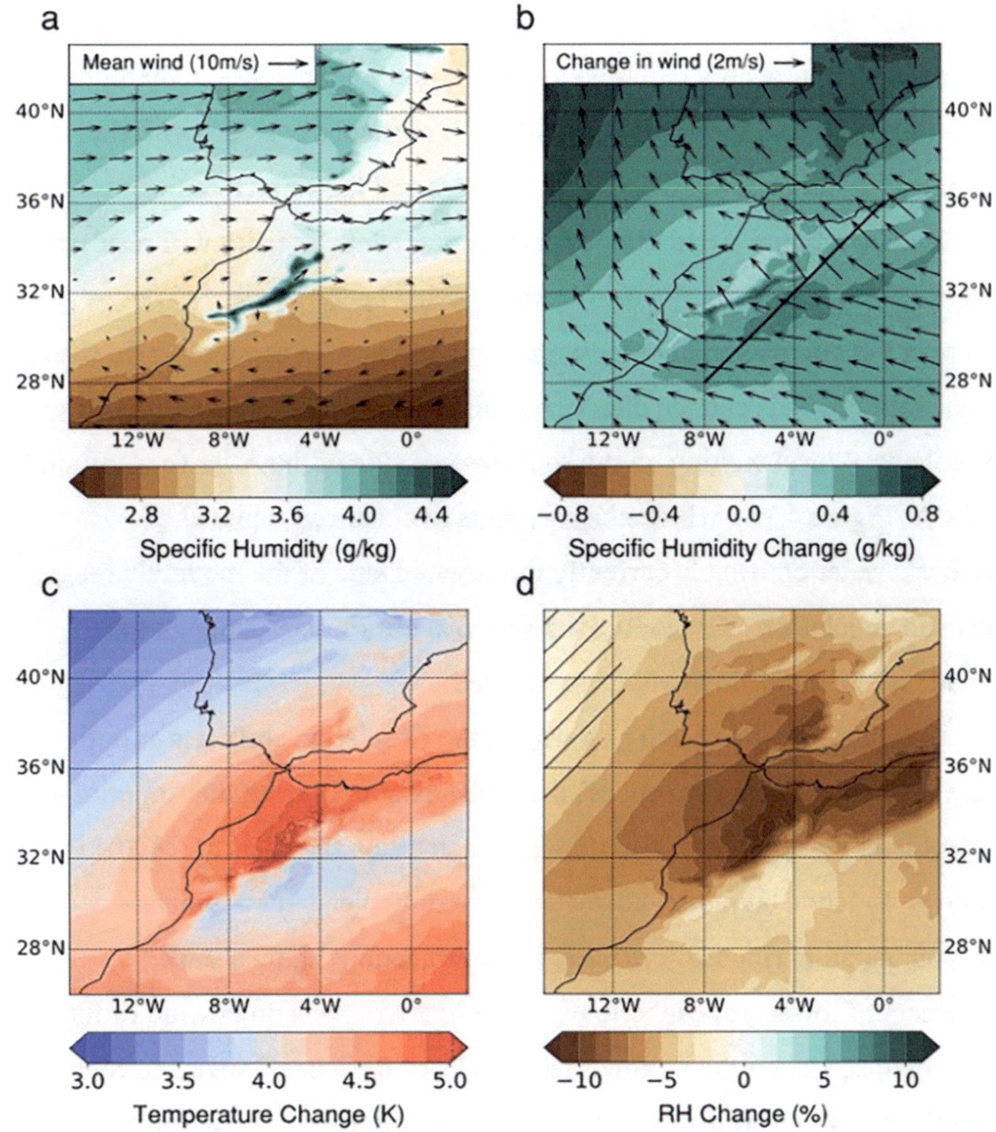

Figure 4.9: a: DJF MRCM ensemble-mean historical (1976–2011) 850 hPa specific humidity (shaded) and circulation (arrows). b, c, d MRCM ensemble-mean change in 850 hPa DJF (b) specific humidity (shaded) and circulation (arrows), (c) temperature and (d) relative humidity. Hatching in d indicates disagreement by one of the three models on the sign of the change. The solid black line in b is used to define a regional wind index (see main text)

wetting trends. The lowest increase is found over the western foothills of the High Atlas. Two factors probably combine to further limit the increase in humidity over this small region. First, the humidity gradient is particularly steep, with the Atlas acting like a barrier between the very dry Sahara and their much moister western flanks. Second, a future circulation with more frequent easterlies will result in strengthened descent of dry air, its moisture having been "squeezed" out as it passed over the mountains. In other words, this region will become a rain shadow as the prevailing winds change direction. This interpretation is further buttressed by the enhanced local wetting on the southeastern side of the Atlas (e.g., around 4°W–32°N, Figure 4.9b), which clearly points to enhanced removal of atmospheric water on what is currently the leeward side of the mountain range. Results and conclusions for the RCP4.5 scenario are similar, although the magnitude of the trends is smaller, as expected.

Therefore, for the region as a whole, the circulation shift acts to both increase dry advection, and to cut off the moisture supply from the Atlantic, consistent with the argument proposed in Chapter 2 on the basis of CMIP5 projections. Relative humidity, however, is not only dictated by moisture content but also temperature (Figure 4.9d). Regional projections show a stark enhancement of lower-tropospheric warming in the regions downwind of future wind anomalies, i.e., on the northwestern flanks of the Atlas Range and Sierra Nevada (Figure 4.9c). At the same time, the warming is relatively weaker, by 1–1.5 °C, on the upslope slide, especially the Saharan Atlas. The geography of the warming and wind change patterns, common to both winter and spring, suggests that topography-driven anomalous vertical motion is an important driver. Strengthened easterlies lead to uplift southeast of the mountain ranges (i.e., upwind) and subsidence on their northwestern slopes (i.e., downwind), respectively associated with adiabatic cooling and warming. Changes in vertical motion near the boundary layer top are consistent with this perspective. In fact, the more realistic topography in MRCM simulations generates stronger subsidence motion and enhanced warming locally (not shown). While temperature gradients across the Atlas Range are negligible in winter, they strengthen over spring as the Sahara Desert warms up fast; consequently, enhanced easterlies likely also contribute to the substantial warming over northwestern Morocco and the Algerian coastline at that time of year. Other factors, such as the soil moisture feedback or the snow feedback over the high mountain peaks, may also play a role (Stéfanon et al. 2014; Walton et al. 2017).

We further explore the role of the regional circulation in determining hydroclimate trends by defining, as a proxy for the regional circulation, a wind index along the African coastline (see line in Figure 4.9b), where future projections show enhanced easterlies. The value of the index is defined as the direction (in degrees) in which the wind blows; for example, a westerly wind will have an index of 0° and a southerly wind an index of + 90°. We focus on the November–April period, the wet season for most of our domain. This index strongly determines to the likelihood of precipitation over Northwestern Morocco and the southern Iberian Peninsula (Figure 4.10-a), a relationship that, overall, MRCM simulations succeed in capturing (Figure 4.10b). Precipitation occurs, as expected, when westerlies prevail (wind index between − 45° and 45°). Similarly, easterlies tend to be associated with dry conditions (index higher than 135° or lower than -135°). On the whole, westerlies are more frequent, as would be expected in the midlatitudes during winter. Consistent with regional circulation

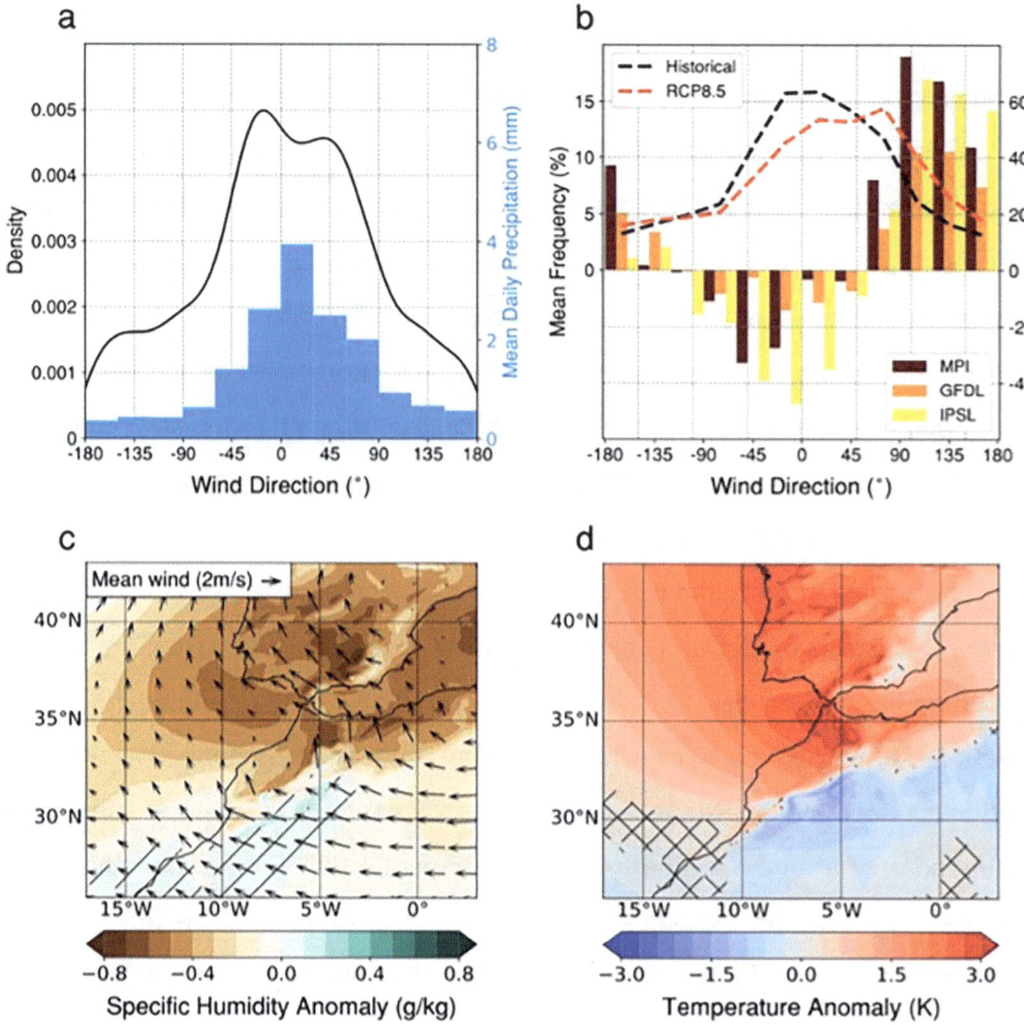

Figure 4.10: a Distribution of November–May daily wind index values from ERA-Interim (solid black line) and average daily precipitation from TRMM over northwestern Morocco and the southern Iberian Peninsula (up to 38°N) in each 30-degree wind direction category (blue bars). b MRCM ensemble-mean distribution of November–May daily wind index values in the historical (dashed black) and RCP8.5 (dashed red) simulations, and relative change in the frequency of each 30-degree wind index category for the three MRCM runs (colored bars). c, d MRCM ensemble-mean average 850 hPa (c) specific humidity anomaly (shaded) and wind values (arrows) and d temperature anomaly, during "southeasterly days" (when the wind index is between + 90° and + 180°) in historical simulation. Hatching in c, d indicates disagreement by one of the three models on the sign of the anomaly

changes (Figure 4.9b), the frequency of north-westerlies decreases across all simulations (by 10–40% across models), while that of south-easterlies goes up significantly. While westerlies are still the most frequent circulation regime, southeasterly winds (index between 90° and 180°) become much more common. Meteorological conditions during southeasterly days are characterized by a substantial atmospheric moisture deficit over the Iberian Peninsula and along the northwestern African coastline (Figure 4.10c), and the wind pattern is very similar to the future projected shift (Figure 4.9b). Additionally, southeasterly days exhibit warmer than- usual temperatures over the same regions with moisture deficit, and cooler-than-usual temperatures on the upwind side of the Atlas and Sierra Nevada, a pattern highly similar to the projected temperature change (Figures 4.9c, 4.10d). This confirms the hypothesis that the circulation shift is responsible for the increase in moisture deficit and enhanced warming seen in regional projections. The combined effects of those mechanisms result in a large decrease in lower-troposphere relative humidity over Northwestern Morocco and Algeria, southern Spain, and the surrounding ocean (Figure 4.9d), from which the precipitation decline follows.

Discussion

Our downscaling results are consistent with the existing literature which highlighted a pronounced wet-season precipitation decline and enhanced summer warming over the southwestern Mediterranean (Diffenbaugh and Giorgi 2012). These trends are consistent with GCM projections of future Mediterranean winter drying as discussed in Chapter 2 (see also Seager et al. 2014; Zappa et al. 2015) and enhanced summertime warming, though GCM deficiencies may cause the latter to be overestimated (Boberg and Christensen 2012). At the local scale, the detailed representation of the complex regional topography and its spatial variability, as well as the resolution of convection, lead to improvements in convective and orographic precipitation (Prein et al. 2015), in keeping with the 0.11° experiments from the EURO-CORDEX and Med-CORDEX initiatives (Jacob et al. 2014, Tramblay et al. 2013, Fantini et al. 2018). Our focus in this study was on winter and spring, when convective precipitation remains limited in the region. MRCM showed pronounced downscaling skill for precipitation during these seasons. Summer precipitation, on the other hand, especially in the mountains, remain challenging to simulate, as it is typically associated with small (< 12 km) convective systems that interact with the topography (Born et al. 2008). New perspectives from convection-permitting simulations could help us refine and better understand trends in convective activity in the warm season (Coppola et al. 2020).

Temperature downscaling also allowed us to refine projections for coastal and mountainous regions, that were unresolved in GCMs. For the coast, where most of Morocco's largest cities are located, mean monthly temperatures may reach extremes of 37°C under the RCP8.5 scenario, a 7°C increase compared to the present-day distribution. Inland cities currently regularly experience mean temperatures well above that threshold, but the relative humidity is much lower there than on the coast, where it remains high even at the peak of summer. This means that summer days may become significantly more uncomfortable in cities like Rabat or Casablanca. This large temperature increase, even under a mitigation scenario, could have significant impacts on Morocco's economy, notably its tourism and agriculture. Higher temperatures—along with increased dryness—in winter, after sowing, and in spring, during grain filling, would certainly impact

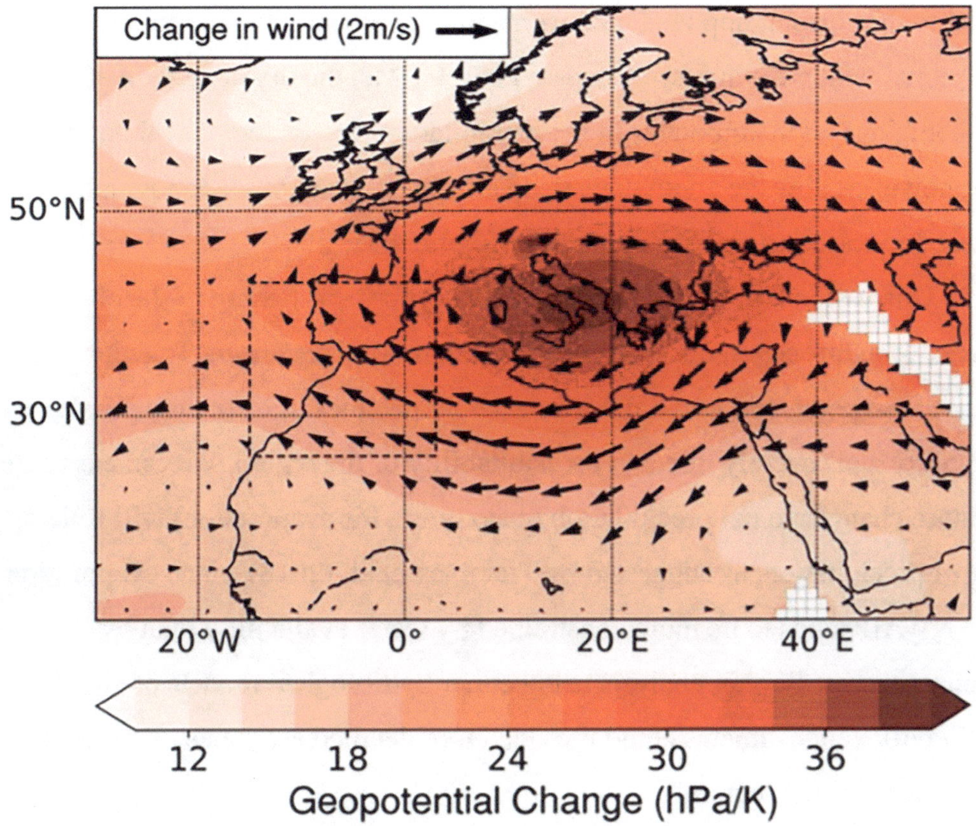

Figure 4.11: CMIP5 multi-model change in DJF 850 hPa geopotential height and wind over the Mediterranean. Our domain region is shown by the dashed black line

cereal yields.

At the regional scale, future projections exhibit dynamical trends characterized by enhanced easterly flow, consistent with basin-wide circulation trends consisting in the development of a strong anticyclone centered on the Mediterranean Basin (see Chapter 2) (Figure 4.11). This pronounced anomalous high, highly robust across GCMs, finds its origin in the combination of large-scale shifts in winter planetary waves due to the strengthening of the mid-latitude jet (Simpson et al. 2016), and to the reduced warming of the Mediterranean Sea with respect to the surrounding land, and its magnitude essentially determines that of the regional winter drying (see Chapter 2). In GCMs, the dry advection effect of this anomalous wind dominates in the precipitation trends; regional simulations show, however, how the anomalous wind also interacts with the topography to enhance the warming and drying pattern (Figure 4.9c, d).

This regional drying is probably the most critical feature of the climate response to anthropogenic forcing in the Western Mediterranean. Together with enhanced temperatures, it will drive substantial declines in river runoff and water availability and severely threaten the habitability of the region. We can expect that the future changes in net precipitation (accounting for evaporation) will follow those of precipitation, as suggested by Tramblay et al. (2018), since evaporation in the NWAIP tends to be more constrained by water availability than by incoming energy. Further analysis using either hydrological models or simple runoff coefficient estimates could provide more detailed estimates of the decline in river flow for the region.

Summary and conclusion

In this study, we sought to project and understand the future wintertime climate of the southwestern Mediterranean, specifically the intense drying that characterizes this region in global climate projections. We performed high-resolution climate downscaling and projections with MRCM driven by the ERA-Interim reanalysis and three carefully-selected GCMs from the CMIP5 archive that correctly represent large-scale feature of winter climate over the NWAIP region. At our 12 km resolution, MRCM showed reliable downscaling skill as tested against observed station- and satellite- based gridded climate products. Projections under the RCP4.5 and RCP8.5 scenarios show a severe precipitation decline over much of the region during the wet season, and strong warming during summer, especially in continental interiors. Local changes in temperature and precipitation are consistent with GCM trends but strongly influenced by the complex topography, and can locally be of higher magnitude, as in the case of warming at high elevations. Temperature and precipitation trends are connected to changes in Mediterranean circulation that translate into enhanced easterly flow over the NWAIP, which, by advecting dry Saharan air and by interacting with the regional orography, lead to a pronounced low-level warming and precipitation decline.

The results of this study underline the high vulnerability of the Western Mediterranean to climate change. This vulnerability is due to its geographical location downwind of a large and very dry desert. In addition, the substantial role played by fine-scale topography and its interaction with regional circulation change make a strong case for the use of high-resolution models in the region. While the selection of GCMs for downscaling could be improved upon, the identified physical mechanisms behind the drying appear robust enough to be found in all regional simulations. Their better understanding that follows from our results may help further constrain the spread of Western Mediterranean future projections.

A version of this article first appeared as

Tuel, A., Kang, S. and Eltahir, E., 2020. Understanding climate change over the southwestern Mediterranean using high-resolution simulations. Climate Dynamics, 56(3-4), pp.985-1001.

Chapter 5: Future projections of High Atlas snowpack and runoff under climate change

Alexandre Tuel
Nabil El Moçayd
Moulay Driss Hasnaoui
Elfatih A. B. Eltahir

The High Atlas is the major source of freshwater for the semi-arid plains of central Morocco. Much of the discharge of the Oum-Er-Rbia and Tensift, the two main rivers of central Morocco, comes from the mountainous terrain where they begin their course. In this region, precipitation essentially falls at elevations above 1000m (Boudhar et al., 2009); below that, it is scarce and evaporation is extremely high, leading to minimal runoff. Though located in a rather warm region, the High Atlas rises up to more than 4000m and often experiences below-freezing conditions between November and March (Boudhar et al., 2009). Consequently, snow is a major component of the regional water cycle (Marchane et al., 2015; Tuel et al., 2020a). It accounts for a substantial fraction of annual runoff, up to 50% in some mountain catchments (Boudhar et al., 2009), and for most of the runoff during spring, as the wet season comes to an end. Snow cover in the High Atlas is characterized by large inter-annual variability (Marchane et al., 2015; Tuel et al., 2020a), mostly following that in wet-season precipitation, itself largely shaped by the North Atlantic Oscillation (Knippertz et al., 2003; Boudhar et al., 2009).

Previous studies indicate that the High Atlas snowpack may be particularly vulnerable to climate change. Climate projections over Morocco – and generally the Mediterranean – agree on robust warming and drying trends under greenhouse gas forcing. By the end of this century, average winter temperatures in the High Atlas could be 2-4°C higher, and precipitation 25-60% lower, depending on the emissions scenario (Ayt Ougougdal et al., 2020; Driouech et al., 2020; Chapter 4). These combined warming and drying trends will unavoidably lead to a snowpack decline. Yet, few studies have analyzed climate change impacts on the local snowpack and regional water availability. López-Moreno et al. (2017) applied a complex physically-based snow model to observed meteorological data at one station in the Moroccan High Atlas, fitted with observed snow depth at the same location. They found that Atlas snowpack was less sensitive to warming and drying than that in other Mediterranean-climate regions (10-15% snow water equivalent decline per degree of warming), because of colder snowpack temperatures associated with high latent heat losses. Still, their results pointed to a decrease in average snow duration of 25-30% and in mean Snow Water Equivalent (SWE) of 30-55% by 2050.

Future trends in runoff in the High Atlas under climate change have also been investigated, notably by Jaw et al. (2015) who analyzed simulations with the Variable Infiltration Capacity model forced by regional climate model output. They found a general tendency to reductions in streamflow, with a strong sensitivity to the forcing model's precipitation trends. Tramblay et al. (2018) took a simple water balance approach, equating long-term net precipitation with water availability, to estimate future changes in dam storage across North Africa. In the High Atlas, they projected a 40-50% decline in water availability under business-as-usual by the end of the 21st century. Only one study tried to quantify the impact of climate change on High Atlas runoff by taking snow dynamics into account: Marchane et al. (2017) developed runoff projections for the Rheyara catchment, south of Marrakech and part of the Tensift watershed, by running conceptual monthly water-balance models incorporating a simple parametric snow module. They projected a 19-63% decline in surface runoff by the middle of the century, dependent on model and scenario. Coupled with population growth, such trends, if realized, will inevitably translate into growing unmet water demand, as shown by Ayt Ougougdal et al. (2020) for the Ourika watershed in the High Atlas.

Thus, while it is clear that the region is headed towards a pronounced decline in snowpack and runoff, much remains to be done to quantify that decline at the catchment level and reduce uncertainties. In this chapter, we hypothesize that a warmer and drier climate will have substantial impacts on the High Atlas snowpack. We develop detailed snowpack projections for the High Atlas under climate change and assess the implications of a declining snowpack on regional runoff. We focus on the Oum-Er-Rbia watershed, a major catchment of the High Atlas. To that end, we apply the methodology of Tuel et al. (2020a) (hereafter T20a), who modeled High Atlas snowpack by applying a simple distributed snow model forced with assimilated remotely-sensed and dynamically-downscaled data. Using satellite-observed snow cover as a baseline for the current climate, we fit and run the snow model with output data from high-resolution regional climate simulations over Morocco as described in Chapter 4. We then quantify the sensitivity of runoff in seven mountain catchments within the Oum-Er-Rbia watershed to large-scale meteorological and snowpack conditions, and use the results to assess the impact of warming, drying and snowpack disappearance on runoff.

Study area and data

Study area

With a length of 550 km, the Oum-Er-Rbia is Morocco's second longest river (Figure 5.1-a). There are about 4 km^3 of available renewable water resources in its basin each year, most of which comes from surface runoff (3.5 km^3), the rest being groundwater (Agence du Bassin Hydraulique de l'Oum-Er-Rbia, 2012). 90% of this water is used to irrigate 350000 hectares of fields, accounting for 30% of Morocco's irrigated land, with the rest supporting the needs of two major cities, Casablanca and Marrakech, and industrial phosphate mining. In addition, mountain runoff is used to generate hydroelectricity. The Oum-Er-Rbia river begins its course in the northeastern portion of the High Atlas, near the city of Khenifra; as it flows westwards towards the Atlantic Ocean, it receives major contributions from northward-flowing tributaries originating in the High Atlas (Figure 5.1-a). Beyond that, the river continues its course in semi-arid plains which bring little additional runoff.

The climate of the area is rather continental, characterized by a large amplitude in the annual cycle of temperature (Knippertz et al., 2003). Minimum temperatures occur in January, when they range from mild (≈12°C) below 1000 m to cold (-5°C) above 3000 m. Temperatures reach their peak in July, at about 35°C below 1000m and 10-15°C above 3000m (Ouatiki et al., 2017). Annual

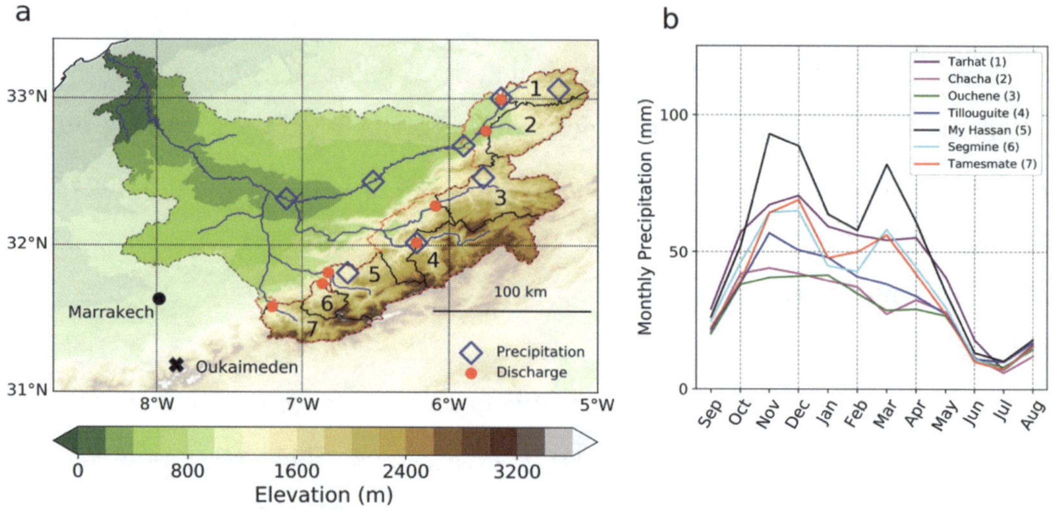

Figure 5.1: (a) Map of study area, the Oum-Er-Rbia watershed, with elevation shown in filled contours. The main waterways are indicated by solid blue lines. Blue diamonds and red circles indicate the location of precipitation and river discharge stations, respectively. The seven catchments defined by the discharge stations are indicated by numbers: (1) Tarhat, (2) Chacha, (3) Ouchene, (4) Tillouguite, (5) Moulay Hassan, (6) Segmine and (7) Tamesmate. The location of the Oukaimeden snow station, outside our study area, is shown by a black cross. (b) Annual cycles of precipitation for the seven catchments, based on TRMM data (1998-2015).

precipitation in the whole basin averages about 400 mm, with a low of 250 mm in the lowland plains, and a high of 800 mm in the mountains to the south (Ouatiki et al., 2017). Most of that precipitation occurs between October and May, when the region is under the influence of North Atlantic westerlies (Knippertz et al., 2003; Tuel and Eltahir, 2018) (Figure 5.1-b). Mountainous areas also experience substantial precipitation during summer, due to small-scale convection (Born et al., 2008). As a consequence, vegetation outside the lower-elevation valleys is sparse, essentially limited to bare soil, grass and occasional shrubs (Baba et al. 2019). Snowfall is common between November and March above 1500m elevation, and it is frequent to observe snow cover persisting for several months above 2500 m (Marchane et al., 2015). Inter-annual variability of snow cover is substantial, however, following that of precipitation (Boudhar et al., 2010). Melt begins in February, and the snowpack is typically gone by the end of May (Tuel et al., 2020a).

We focus specifically on the 13,610 km^2 domain analyzed by T20a (Fig. 5.1-a), which encompasses most of the area above 1000 m altitude within the Oum-Er-Rbia watershed and thus most regions that receive significant snowfall (Marchane et al., 2015). Elevation in this domain ranges from 621 m to 3890 m. with an average of 1882m.

Hydroclimatological data

We use for this study a mixture of model-, station- and satellite-based hydrometeorological data. Model-based data is described in the next section. Daily precipitation data are available at seven stations in the study area, including three at more than 1200m elevation, over the 1980-2015 period (Figure 5.1-a) (see also Table 5.1). For each precipitation series, we conduct basic quality checks following Durre et al. (2010). We then discard all the months for which more than 10% of the data is missing or flagged. This leaves more than 95% of the data for analysis. This data is used to check the validity of the satellite precipitation products described below. In addition, daily discharge measurements are available at seven locations as well, between 1978 and 2015. Each has at most 0.5% of missing data. We implement a simple quality control following Gudmundsson et al. (2018). Days with negative discharge are flagged, as well as all consecutive periods of more than 10 days during which discharge values are equal and larger than zero. We also flag as suspect daily discharge values Q such that log(Q+0.01) is more than six standard deviations away from its mean value, with mean and standard deviation computed over a 10-day period around the corresponding calendar day over the whole time series.

The locations of the discharge stations define seven sub-catchments for which runoff will be analysed, from north to south: Tarhat, Chacha, Ouchene, Tillouguite, Moulay Hassan, Segmine and Tamesmate (Figure 5.1-a). Area and average elevation for each catchment are given in Table 5.2. Their average elevation varies from 1460 to 2360m. We remove the contribution from base flow by subtracting the minimum monthly discharge value for each catchment and each

Data	Description	Availability
Station precipitation	Daily precipitation measured at seven locations in the Oum-Er-Rbia watershed	1980-2015
Station discharge	Daily discharge measured at seven locations in the Oum-Er-Rbia watershed	1978-2015
TRMM TMPA 3B42 version 7	Satellite-based 3-hourly precipitation at 0.25° resolution	1998-present
CHIRPS v2.0	Satellite- and station-based 6-hourly precipitation at 0.05° resolution	1981-present
MODIS Land Surface Temperature L3 version 6 (MOD11A1)	Satellite-based land surface temperature at 1 km resolution	2000-present
MODIS Terra snow cover daily L3 (MOD10A1)	Satellite-based fractional snow cover at 500 m resolution	2000-present
ERA/MRCM	Regional downscaling of ERA-Interim with MRCM at 12 km resolution (from Tuel et al. (2020b))	1981-2011
ERA/GCM	Regional downscaling of three CMIP5 GCMs (IPSL-CM5A-LR, GFDL-ESM2M and MPI-ESM-MR) with MRCM at 12 km resolution (from Tuel et al. (2020b))	1976-2005 (historical) and 2071-2100 (RCP4.5 and RCP8.5)
SRTM 90-meter resolution version 4.1 (STRM90)	Satellite-based elevation at 90 m resolution	N/A

Table 5.1: Overview of the datasets used in this study.

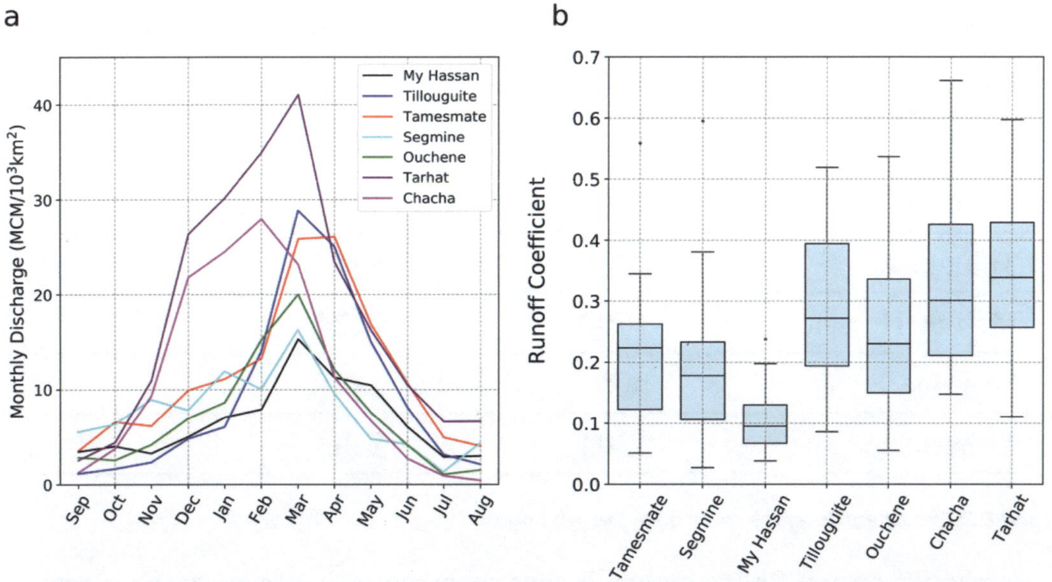

Figure 5.2: (a) Annual cycles of monthly runoff at the seven runoff gauges, after base flow removal and normalization by catchment area (km2). (b) Boxplot of annual runoff coefficients for the seven catchments (1982-2011), using ERA/MRCM precipitation bias-corrected with TRMM data.

hydrological year (September-August). This correction is minor for all catchments except Tarhat, the northernmost one, which includes the headwaters of the Oum-Er-Rbia river, and receives a substantial contribution of base flow to its annual discharge. In particular, the flow at Tarhat remains high during summer (≈ 35% of its wet-season peak), likely due to groundwater discharge from deep mountain aquifers. Annual cycles of corrected monthly discharge are shown on Figure 5.2-a.

Satellite-based data is used for basin-wide precipitation, temperature and snow cover. 3-hourly precipitation from the TRMM TMPA (TRMM Multi-Satellite Precipitation Analysis) 3B42 version 7 dataset is used as the reference

Name	Area (km^2)	Mean elevation (m)
Tarhat	997	1627
Chacha	1519	1460
Ouchene	2391	1953
Tillouguite	2488	2363
Moulay Hassan	1700	2124
Segmine	506	1897
Tamesmate	1303	2198

Table 5.2: Characteristics of the seven analyzed sub-basins

precipitation dataset for the region. It consists in remotely-sensed data corrected with rain gauge data on a monthly basis (Huffman et al., 2007). The data cover the period 1998 to present, at a resolution of 0.25°. While satellite-based precipitation data suffers from numerous biases (Milewski et al., 2015; Derin et al., 2016; Hashemi et al., 2017), it is often the only option available in complex terrain where stations are scarce. Milewski et al. (2015) and Ouatiki et al. (2017) assessed the accuracy of the TRMM 3B42 V7 dataset in the Oum-Er-Rbia and found that, although unreliable at the daily timescale, it offered satisfactory estimates of precipitation if averaged in space or time. Annual cycles of TRMM precipitation for the seven catchments are shown on Fig.1-b.

For comparison, we also consider the CHIRPS dataset, available from 1981 onwards at a 0.05° resolution (Funk 2015). CHIRPS is produced by combining high-resolution satellite-based precipitation with station and fine-scale topography data. In our region of focus, TRMM and CHIRPS show some differences (Figure 5.3-a,b): CHIRPS is notably wetter, particularly near the Tizi

N'isly station. A comparison of monthly values with the four available stations above 1000m suggests a rather dry bias in TRMM and inconsistent biases in CHIRPS (Figure 5.3-c,d). Both datasets have a dry bias over the north-eastern corner of our domain (around the Kenifra station). Across the four stations shown on Figure 5.3, absolute biases range from 13.4 mm to 25.6 mm for TRMM and from 18.6 mm to 25.2 mm for CHIRPS. Our purpose here is not to perform an in-depth comparison of the performance of the two datasets, but to test the robustness of the runoff projection results to variability in reference precipitation. To bias-correct regional climate model output for snow modeling (see Snow odel section in Methods, later in this chapter), however, we follow T20a and consider only the TRMM dataset.

Reference surface air temperature is derived from MODIS Land Surface Temperature (LST) product MOD11A1 L3 version 6 at 1km resolution (Wan, Z., Hook, S., Hulley, 2015). We refer to T20a for details on data filling and correction. Observed snow cover area for the region is extracted from the MODIS Terra snow cover daily L3 product (MOD10A1) at 500m resolution (Hall and Riggs, 2016). Snow cover is detected using values of the Normalized-Difference Snow Index, based on reflectances in the visible/near infrared and middle infrared. We apply the correction methodology described in Marchane et al. (2015) which allows to substantially reduce the number of missing data points due, mainly, to cloud cover, and average the data at the weekly timescale as in T20a. We refer to these two studies for a discussion of the accuracy of the MODIS dataset in this region. All MODIS data is available from February 2000 to present. Elevation data is taken from the Shuttle Radar Topography Mission 90-meter resolution dataset version 4.1 (STRM90) (Jarvis et al., 2008), and

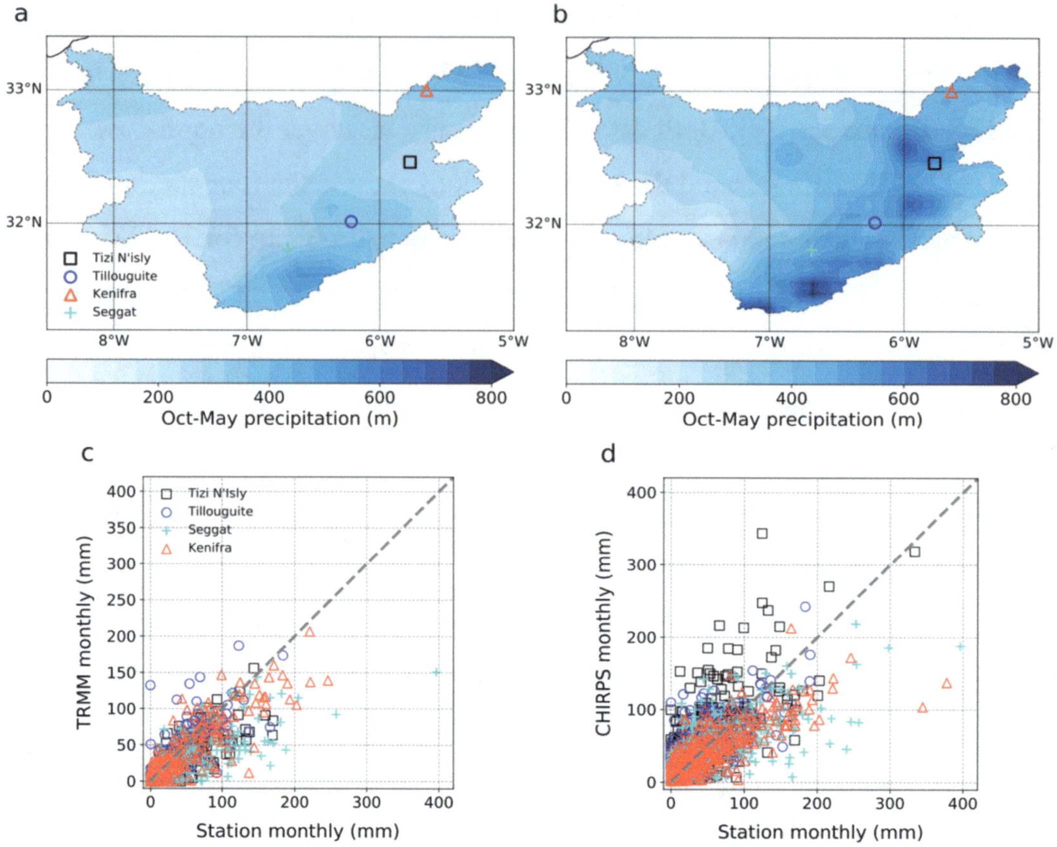

Figure 5.3: (a) October-May average precipitation (mm) over the Oum-Er-Rbia watershed, from TRMM (1998-2015). The four available precipitation stations above 1000m elevation are indicated by symbols. (c) Monthly precipitation at the four stations shown on (a) against corresponding TRMM values. (b,d) Same as (a,c) but for the CHIRPS dataset (1981-2015).

interpolated to the approximately 1km resolution of the MODIS land surface temperature data.

Regional climate simulations

We consider the regional climate downscaling data and projections developed in Chapter 4 for the Western Mediterranean, at a 12km resolution, using the MIT Regional Climate Model (MRCM). MRCM is based on the Abdus Salam International Centre for Theoretical Physics Regional Climate Model

Version 3 (RegCM3) (Pal et al., 2007), but with significant enhancements of model physics, and notably a coupling with the Integrated Biosphere Simulator land surface scheme (IBIS). Dynamical downscaling is performed for ERA-Interim (1982-2011) (Dee et al., 2011) as well as three carefully selected GCMs from the Coupled Model Inter-comparison Project Phase 5 (CMIP5) (Taylor et al., 2012): MPI-ESM-MR, GFDL-ESM2M and IPSL-CM5A-LR, for the historical (1976-2005) and RCP4.5 and 8.5 (2071-2100) scenarios. Details of the simulations, including model setup and performance, can be found in Chapter 4.

Methods

The methodological framework adopted in this study is summarized in Fig. 4. We start from GCM simulations from the CMIP5 archive, dynamically-downscaled with MRCM. We apply bias-correction to the MRCM output, using observed and ERA-Interim data downscaled with MRCM as target, which we then feed into a 1 km resolution snow model over our study region to reconstruct snowpack under the various emissions scenarios. Finally, a statistical model is developed for catchment runoff coefficients (RCs), in order to make projections of runoff under future climate conditions.

Bias-correction and statistical downscaling

6-hourly wind speed, specific humidity, air temperature, precipitation, and downward longwave and shortwave are extracted from the MRCM output over our domain. We then bias-correct MRCM output when reliable observations are available, and downscale it to the MODIS 1km resolution at which we run the snow model. Air temperature and precipitation in all MRCM runs (ERA-Interim and GCM-driven simulations) are bias-corrected at the 6-hourly timescale using the MODIS LST-derived air temperature and TRMM precipitation at their native resolutions as respective targets, via the CDF-transform method (Michelangeli et al., 2009). Alone among the three GCMs, the IPSL-CM5A-LR model exhibits a negative bias in wet days that we correct at each grid cell by randomly generating wet days of magnitude drawn from the corresponding distribution of wet-day precipitation in the TRMM dataset. For bias correction, reference periods for "perfect" observations are 1998-2011 for TRMM and 2000-2011 for MODIS. The corresponding periods in the simulations are the same for ERA/MRCM, and the 1992-2005 and 1994-2005 periods, respectively, for each of the GCM-driven simulations. All bias corrections are performed for the cold (November-April) and warm (May-October) seasons separately. As to other variables, for which reliable observations are not available, the output from the ERA-Interim run (hereafter referred to as ERA/MRCM) is taken as "ground truth". Therefore, wind speed, downward long- and shortwave radiation and specific humidity in the GCM runs is bias-corrected against the ERA/MRCM values over the 1982-2005 reference period. The downscaling to the MODIS 1km resolution is then obtained as follows. Precipitation, downward long- and shortwave and wind are simply downscaled with no elevation correction, keeping the 12km-grid value for all the 1km grid cells. Specific humidity, however, is downscaled with a lapse-rate μ estimated empirically from the simulations at each time step:

$$\log(q) = \log(q_{12}) + \mu \cdot (z - z_{12}) \qquad (1)$$

where q_{12} is the specific humidity in a given 12-km resolution grid cell of elevation z_{12}, and q the downscaled value at elevation z. No additional downscaling for temperature data is required since the downscaling is embedded

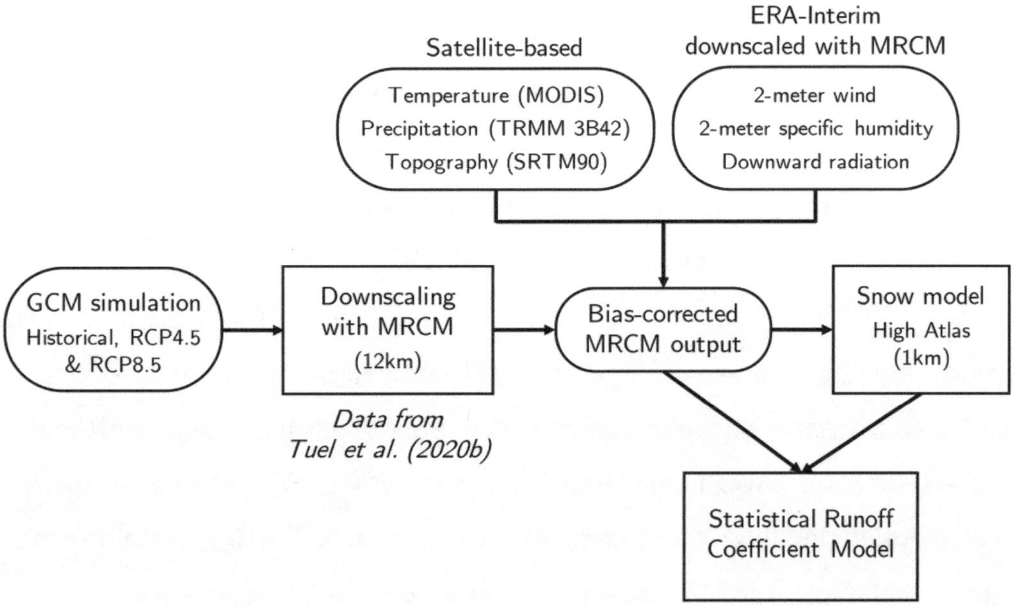

Figure 5.4: Summary of methodology to assess climate change impacts on snowfall, snowpack and runoff in the Oum-Er-Rbia watershed.

into the bias-correction step. The target MODIS LST-derived air temperature data in the bias-correction is indeed already at the required 1km resolution.

Snow model

We apply the SNOW-17 model (Anderson, 2006) with a radiation-derived temperature index for melt (Follum et al., 2015) as described in T20a. SNOW-17 simulates snow accumulation and loss based on meteorological variables, and accounts for the various energy balance equation terms. Snowpack is characterized by its snow water equivalent (SWE) and heat deficit, defined as the amount of heat (in equivalent mm of SWE) required to bring its temperature up to freezing point. We also integrate the bulk-aerodynamic formulation of sublimation detailed in T20a. Readers are referred to Follum et al. (2015) and Anderson (2006) for more details about SNOW-17.

We run the model at a 6-hourly time step and at the native MODIS LST 1km resolution over the domain shown on Figure 5.1-a. Model output includes SWE, as well as melt and sublimation fluxes at each time step. Following T20a and Boudhar et al. (2011), SWE given by the model is then translated into snow cover fraction for each grid cell using the following relationship:

$$SC = 0.85 \times \tanh(k \cdot SWE) \qquad (2)$$

with $k = 100$. We select 0.85 as the maximum allowable snow-covered fraction because snow fraction values larger than 85% are extremely rare in the 500m MODIS data, even at high elevations. This likely reflects the strong small-scale variability of snow cover in the High Atlas (Baba et al., 2019). The snow model requires optimizing three parameters: M_f (melt factor), NMF_{max} (maximum negative melt factor) and $TIPM$ (coefficient used in updating snowpack temperature) (Anderson, 2006). In keeping with T20a, parameter calibration is performed by maximizing the Nash-Sutcliffe coefficient (Nash and Sutcliffe, 1970) between the annual cycles of observed (MODIS) and simulated snow cover at 250 randomly selected grid points within the snow domain. We force the elevation distribution of these 250 points to match that of the whole domain. The annual cycles are computed for the 1995-2005 period in the GCM-driven simulations, 2000-2011 period in the ERA-Interim simulation and 2000-2011 period in the MODIS series. Parameters are calibrated independently for each of the simulations (ERA-Interim and three GCMs) in their respective reference periods. For the future simulations, parameter values are kept constant, equal to their calibrated values.

Statistical modeling of runoff coefficients

We model catchment runoff coefficients (RCs), defined as total October-May discharge divided by total October-May precipitation, across time and space as functions of large-scale hydrological variables by adopting a panel regression framework. Panel regression allows to enhance the effective size of a dataset and to obtain more robust estimates of the response to selected covariates compared to more traditional regression approaches (Steinschneider et al., 2013; Davenport et al., 2020). It also allows to account for static (space-dependent) and time-varying (time-dependent) factors, although here, with only seven catchments, we do not have enough data to make robust statements about static factors responsible for the disparity in average RC (Figure 5.2-b). Therefore, we focus on time-varying covariates, and consider a fixed-effects formulation:

$$\log\left(RC_{j,t}\right) = \log\left(\overline{RC_j}\right) + \sum_i \beta_i X_{j,t}^i + \epsilon_{j,t} \quad (3)$$

where $j \in \{1, \ldots, 7\}$ is the catchment index, t is the time index, i is the covariate index, and $\overline{RC_j}$ represents time-invariant, watershed-specific fixed effects (drainage area, land cover, mean climate), $X_{j,t}^i$ are the covariates, β_i are regression coefficients and $\epsilon_{j,t}$ is random noise. For covariates, we consider catchment-averaged October-May precipitation (P), relative humidity (RH), temperature (T), snow water equivalent (SWE) and snow fraction of precipitation (SF). Runoff has indeed been shown to be sensitive to these variables (e.g., Berghuijs et al., 2017). Enhanced precipitation or relative humidity lead to wetter soils and can be expected to be associated with higher RC values. Similarly, higher temperatures increase evapotranspiration and tend to decrease runoff. Finally, increased snow cover favors losses by sublimation and shifts the distribution of runoff regimes towards slower runoff as opposed to rapid overland flow following rain storms.

However, larger snow cover may also increase the risk of rain-over-ice events, which tend to have very high runoff coefficients (Davenport et al., 2020). Environmental model covariates are calculated using the ERA/MRCM run and associated snow model output. Runoff coefficients (RCs) for each catchment are defined as observed total October-June discharge divided by total October-June precipitation over the catchment. Precipitation is obtained from the ERA/MRCM run bias-corrected with TRMM. To assess the robustness of the results to the choice of TRMM to bias-correct precipitation, we also calculate RCs based on precipitation series bias-corrected with the CHIRPS dataset (available from 1981-present).

Model selection is performed by stepwise regression: starting from a model with no covariates, covariates are added one at a time in the order of highest improvement to model skill, as determined by its Akaike Information Criterion (AIC). At each step, we also test whether removing any of the currently selected variables and replacing it by one of the remaining, non-selected ones, brings any improvement. To estimate the sensitivity of RC to changes in climate conditions, we modify covariate values in the 1982-2011 ERA-Interim downscaled simulation by adding projected long-term changes in the GCM-driven simulations:

$$\log(\widehat{RC}_{j,t}^{m}) = \log(\overline{RC}_j) + \sum_{i \in I} \beta_i (X_{j,t}^{i} + \bar{X}_m^{i}) \qquad (4)$$

where $m \in \{1,2,3\}$ is model index, I is the set of optimal covariates, $X_{j,t}^{i}$ are ERA-Interim downscaled covariate values and $\bar{X}_m^{i}$ represent long-term covariate changes drawn at random according to:

$$\bar{X}_m^i \sim N\left(\mu_{m,\mathrm{rcp}}^i - \mu_{m,\mathrm{hist}}^i, \sqrt{\sigma_{m,\mathrm{rcp}}^i + \sigma_{m,\mathrm{hist}}^i}\right) \quad (5)$$

where $\mu_{m,s}^i$ (respectively $\sigma_{m,s}^i$ is the average (respectively standard deviation) of covariate *i* in model *m* and scenario *s*. Results for the three models are then pooled together to yield a future distribution for $RC_{j,t}$.

Results and discussion

Snowpack modeling and projections

Annual cycles of reconstructed snow cover as a function of elevation are shown on Figure 5.5. Overall, all models succeed in accurately reproducing snow cover dynamics in the region, although the ERA-Interim simulation tends to have a positive bias at high elevations, particularly above 3000m (Figure 5.5-b) and all simulations have a negative bias at low elevations (Figs. 5-d,e and 6-a,b). The GCM-driven experiments generally show too little snow cover, and a later snowpack build-up (Figure 5.5-f). December-March average snow cover over the whole modeled area reaches 1460 km^2 in MODIS observations but only 1275 km^2 in the ERA-Interim driven run and 1185 km^2 in the ensemble mean historical average – a bias largely concentrated at low elevations. Still, except for elevations below 2000 m, mean snow cover mostly remains close (±20%) to MODIS values. The intermodel spread in GCM-driven simulations also generally overlaps with observed snow cover values. Consistent with these results, the spatial distribution of snow-to-precipitation ratio is generally well-reproduced in the GCM-driven experiments (Figures 5.7-a,d and A1), apart from negatives biases at low elevations associated with the underestimation of snow cover (Figure 6). Inter-annual variability in basin-wide snow cover is lower in the simulations compared to MODIS (standard deviation of 220-440 km^2 compared to 480 km^2 in MODIS) but the discrepancy is mainly due to the negative bias at low elevations, where snow plays a much more limited role in the overall water balance (T20a). Still, in the current climate, we find that annual relative sublimation losses are strongly linked to annual-mean precipitation, a relationship that the GCM experiments are able to capture (Figure 5.8-a). Losses from latent heat fluxes are much smaller during wet years as compared to dry years, a relationship robust across all experiments. Wet years indeed bring higher RH over the region, due to enhanced moisture advection from the Atlantic which more than compensates for the larger heat advection and increased air temperatures that also occur in parallel (Knippertz et al., 2003). All experiments also agree reasonably well on the evolution of sublimation losses with elevation (Fig. 8-b). The MPI-ESM-MR and IPSL-CM5A-LR simulations exhibit a slight negative bias, but which remains within the uncertainty range in the assimilated run (T20a).

Future projections show a stark decline in snow cover across all the region (Figures 5.6 and 5.9). The greatest relative decline is at low elevations, as expected since they are already seldom above the 0°C line (Boudhar et al., 2016). Above 2500 m, projections still exhibit a 30-40% decrease in snow cover area under RCP4.5 and 50-60% decrease under RCP8.5. Maximum snow cover extent is reached around the beginning of February in all scenarios. However, snowpack build-up and melt occur respectively later and earlier in the season. The projected decline is even steeper in terms of snowpack water content (Figure 5.10): SWE is reduced on average by 60% under RCP4.5 and 80-85% under RCP8.5, bringing peak SWE value from about 125 million m^3 (MCM) down to 20 MCM (Figures 5.10-f, 5.11). Inter-model spread is largest in the RCP4.5 simulations, in which peak SWE ranges from 25-60% of its historical value.

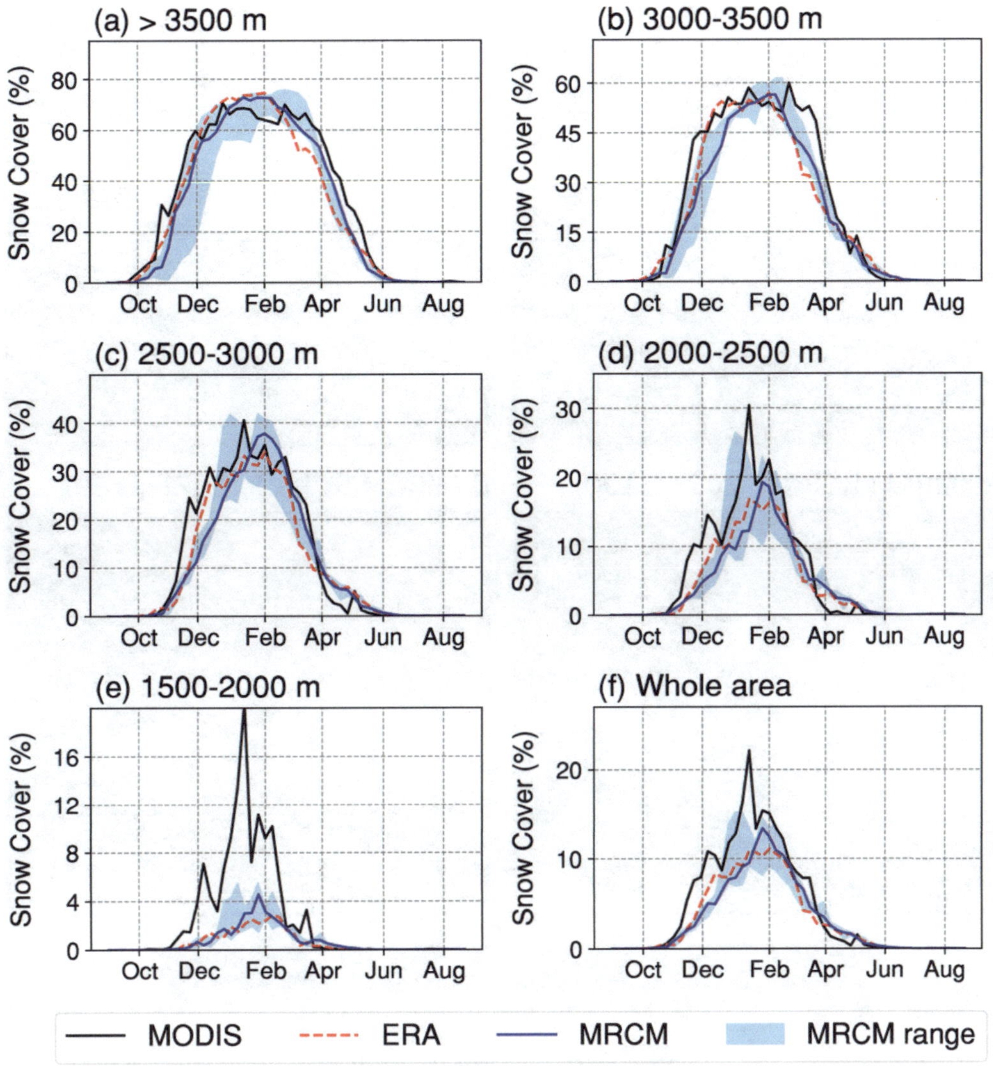

Figure 5.5: Annual cycles of snow cover (in %) in the MODIS observations (2000-2011; black), ERA- Interim simulation (2000-2011; dashed red) and three GCM-driven historical simulations (1994-2005; solid blue: median; blue shading: 3-model range), at various elevations ranges within our study area: (a) > 3500m, (b) 3000-3500m, (c) 2500-3000m, (d) 2000-2500m, (e) 1500-2000m and (f) whole area.

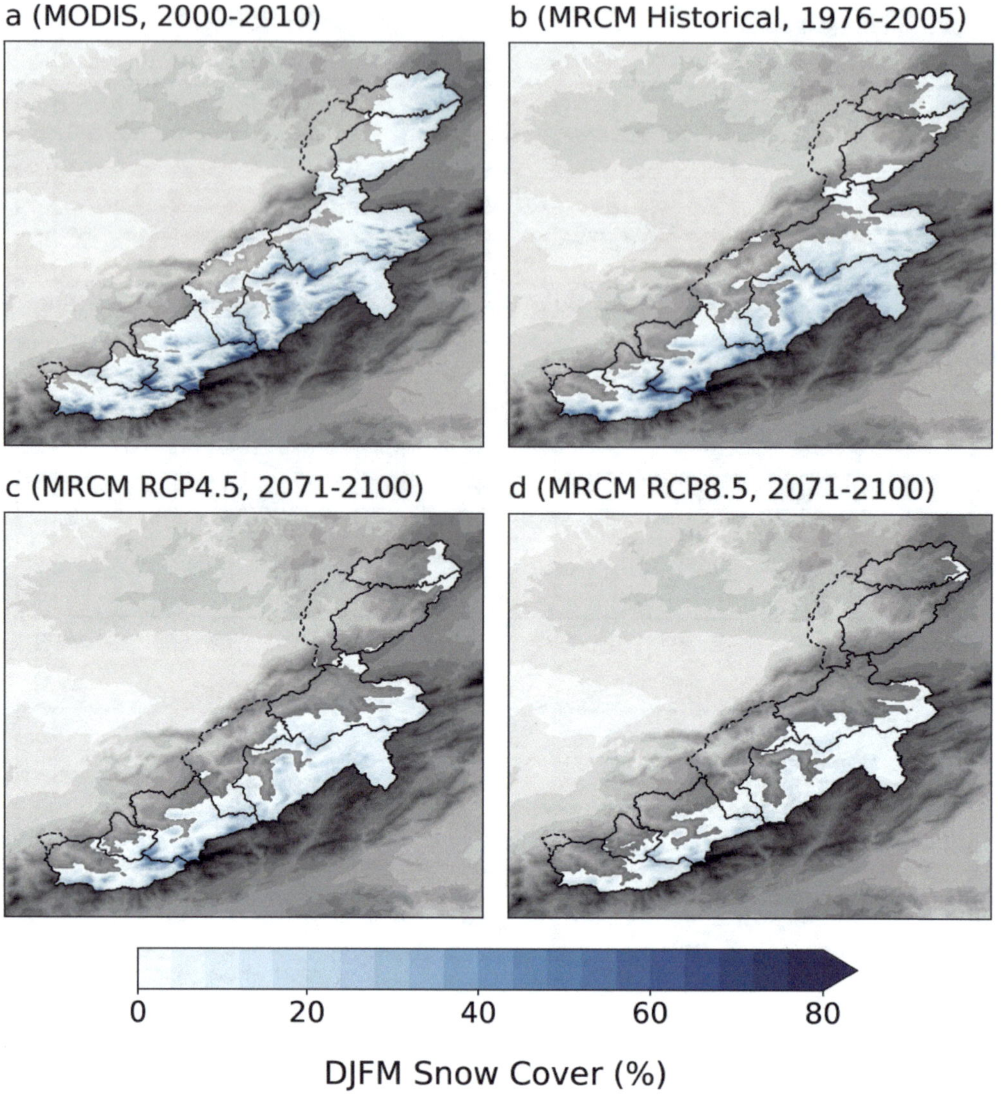

Figure 5.6: Mean December-to-March (DJFM) fractional snow cover (%) over the basin in (a) MODIS (2000-2010) data, and (b-d) three-GCM average under the (b) historical (1976-2005), (c) RCP4.5 (2071-2100) and (d) RCP8.5 (2071-2100) experiments.

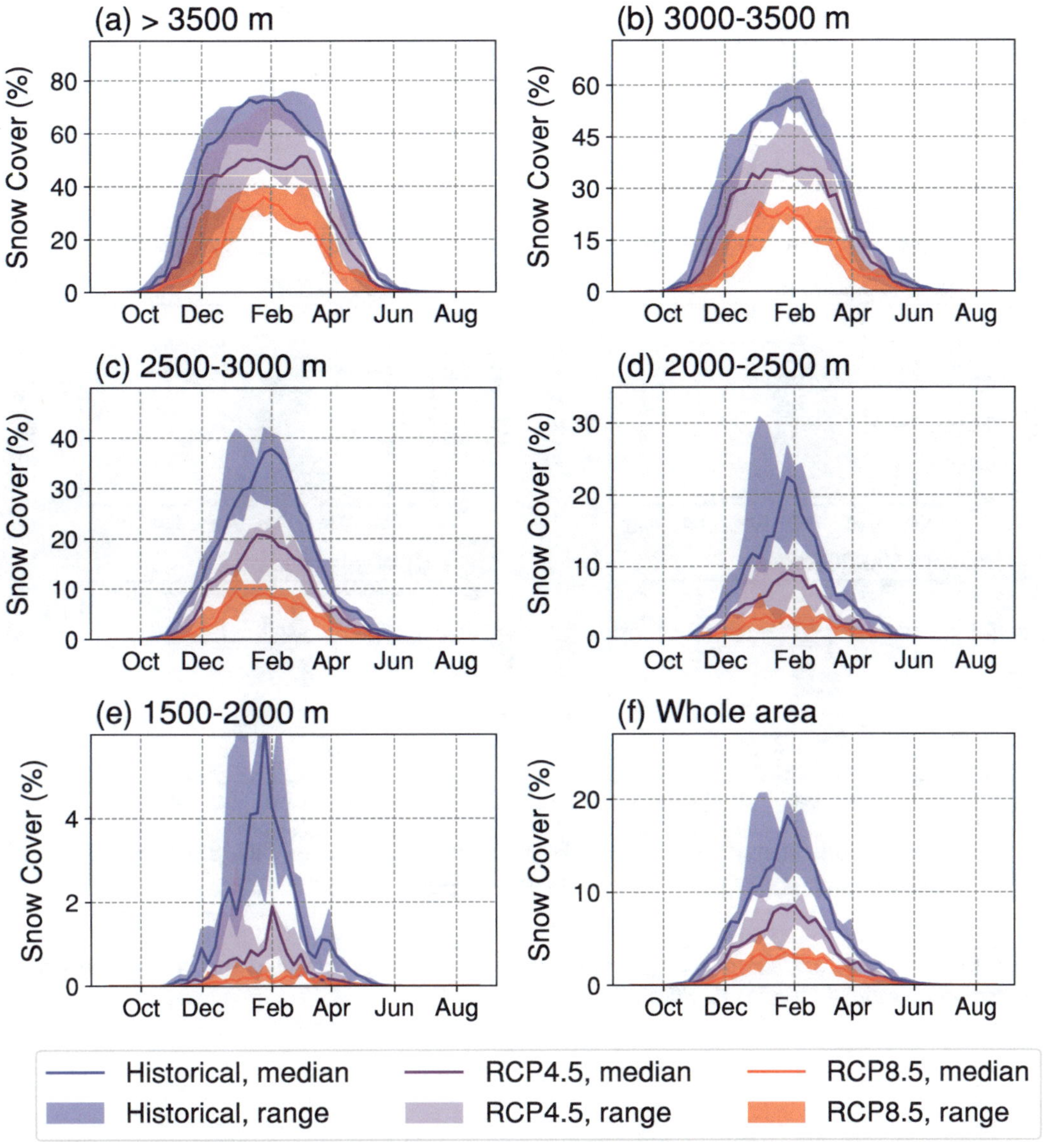

Figure 5.9: Annual cycles of snow cover (in %) in the three GCM-driven experiments under the historical (blue, 1976-2005), RCP4.5 (purple, 2071-2100) and RCP8.5 (red, 2071-2100) scenarios, at various elevations ranges within our study area: (a) > 3500m, (b) 3000-3500m, (c) 2500-3000m, (d) 2000-2500m, (e) 1500-2000m and (f) whole area. Solid lines represent the three-model medians and the shading corresponds to the three-model spreads.

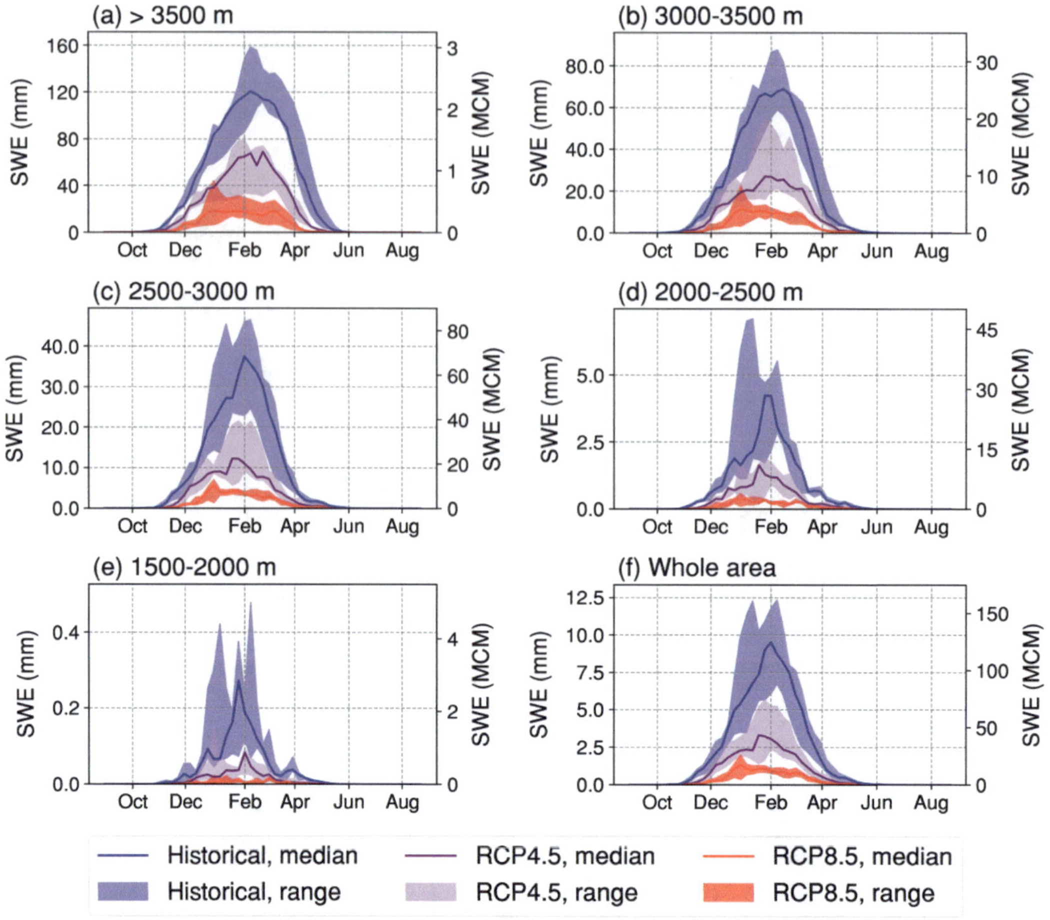

Figure 5.10: Same as Fig. 5.9, but for average snow water equivalent (mm, left-hand axis) and corresponding total snow water content (million m³, MCM).

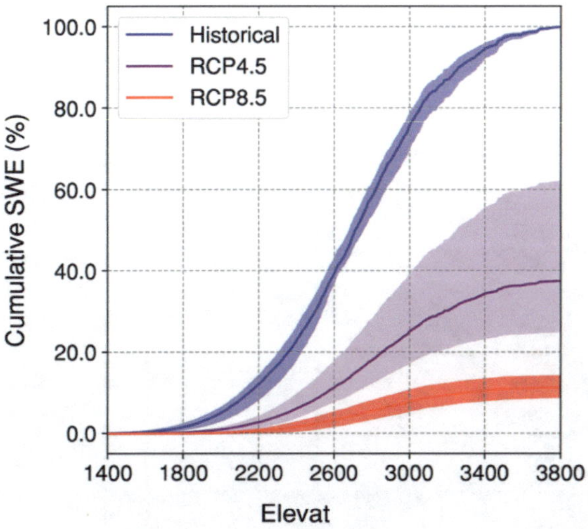

Figure 5.11: Distribution of cumulative basin-wide SWE with elevation in the GCM-driven experiments, under the historical (black), RCP4.5 (purple) and RCP8.5 (red) scenarios. SWE is normalized in each model by that model's historical total basin-wide SWE.

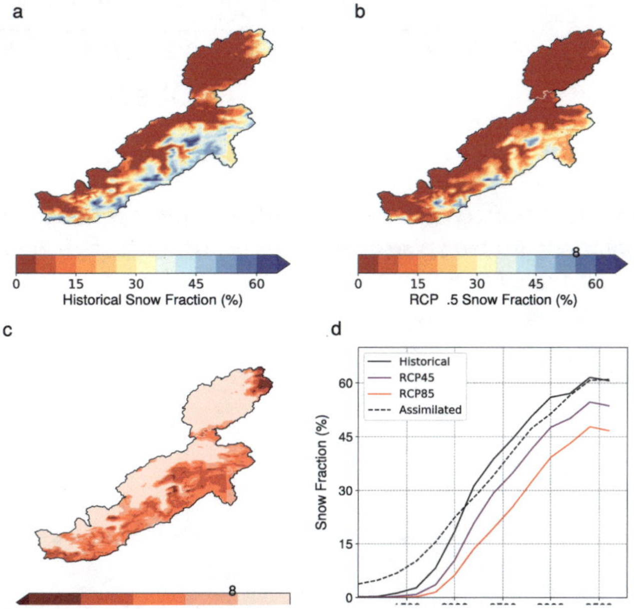

Figure 5.7: (a-b) Snow fraction of annual precipitation in the (a) historical and (b) RCP8.5 scenario (average between all three GCM-driven simulations). (c) Difference between (b) and (a). (d) Snow fraction of annual precipitation as a function of elevation, in each scenario (historical, RCP4.5 and RCP8.5; three-model average) and in the assimilated control simulation of T20a.

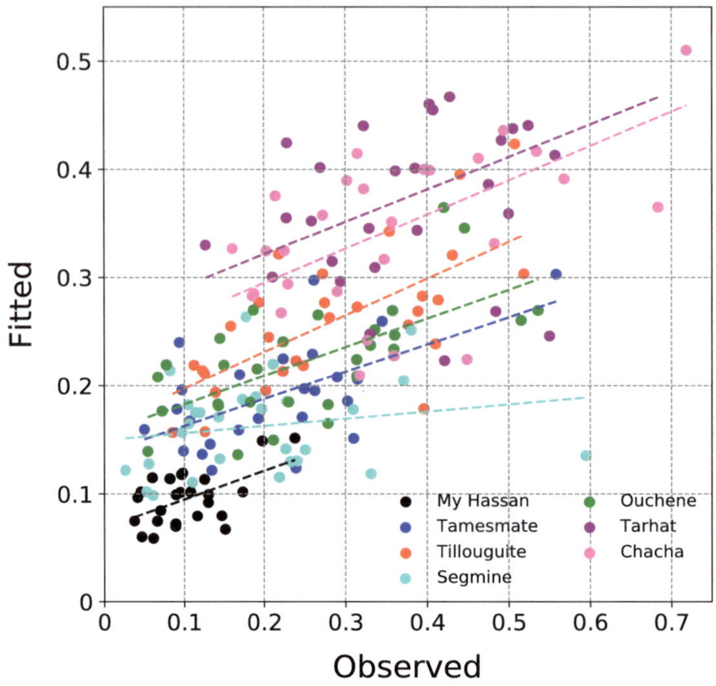

Figure 5.12: Fitted runoff coefficient values against observed values (defined with TRMM precipitation), for the seven catchments in our study area. Best-fit linear regression lines are shown by dashed lines.

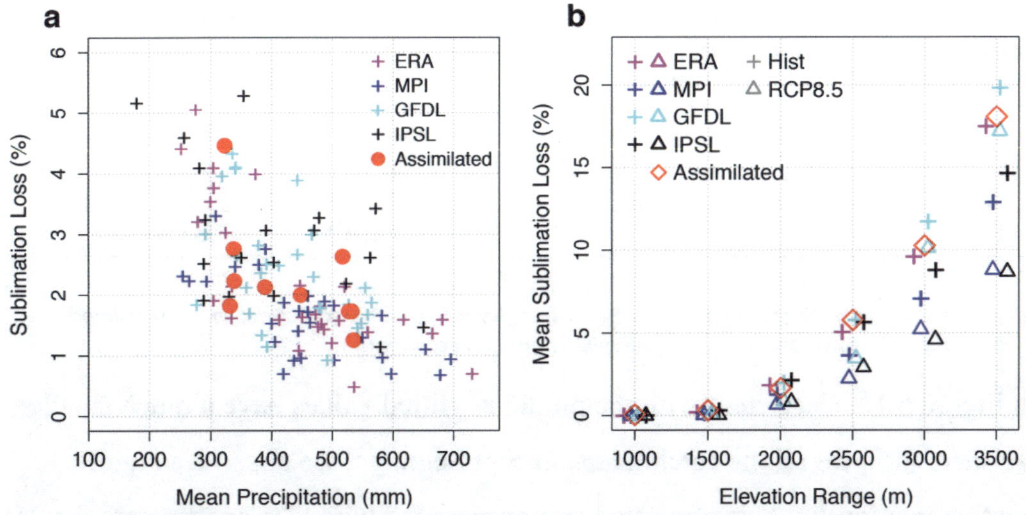

Figure 5.8: (a) Fraction of annual precipitation lost by sublimation against total annual precipitation in the MRCM downscaled experiments forced with ERA-Interim (1982-2011, magenta), the three GCMs (1976-2005): MPI-ESM-MR (blue), GFDL-ESM2M (cyan) and IPSLCM5A-LR (black), and the assimilated snow run forced with MODIS and TRMM data only (red, data from T20a). (b) Fraction of annual precipitation lost by sublimation as a function of altitude range in our study area, for the assimilated run from T20a and in the various downscaled experiments, for historical (1976-2005, "+") and RCP8.5 (2071-2100, "Δ") scenarios.

Runoff modeling and projections

The panel regression and model selection framework are applied to runoff coefficients and selected covariates over the 1982- 2011 period. Stepwise regression yields as optimal covariates relative humidity and snow fraction. The adjusted r^2 is equal to 0.30, meaning that these two covariates explain a small third of inter-annual variability in RC. Fitted RCs against observations are shown

	MRCM-BC		CHIRPS	
r^2	0.30		0.36	
SF	-1.84	4E-03	-1.47	2E-02
RH	5.71	2E-05	8.71	1E-09
Pr	–	–	2.04E-03	1E-06

Table 5.3: Runoff coefficient model results. The bottom three lines show coefficient values (left-hand column) and their statistical significance (p-value, right-hand column).

on Figure 5.12. Consistent with the model r^2, fitted values have a much smaller variance. Still, for all the catchments, except Segmine, we observe a significant positive relationship between fitted and observed values. The coefficients for RH and snow fraction are both significant, with $\beta_{RH} > 0$ and $\beta_{SF} < 0$ (Table 5.3). In other words, all else being equal, a larger RH yields more runoff and more of the precipitation falling as snow yields less runoff. These results are robust to the choice of the precipitation dataset. When using the CHIRPS dataset, average RCs are generally lower, due to CHIRPS's wet bias compared to TRMM (Figure 5.3); the optimal RC model includes RH and SF as well, but also precipitation as a third variable (Table 5.3). The values of β_{RH} and β_{SF} obtained with CHIRPS data are similar, although β_{SF} is slightly less significant. In addition, for both precipitation datasets, the effect of RH on runoff coefficient, as measured by the respective regression coefficient β_{RH}, is 3-5 times that of snow.

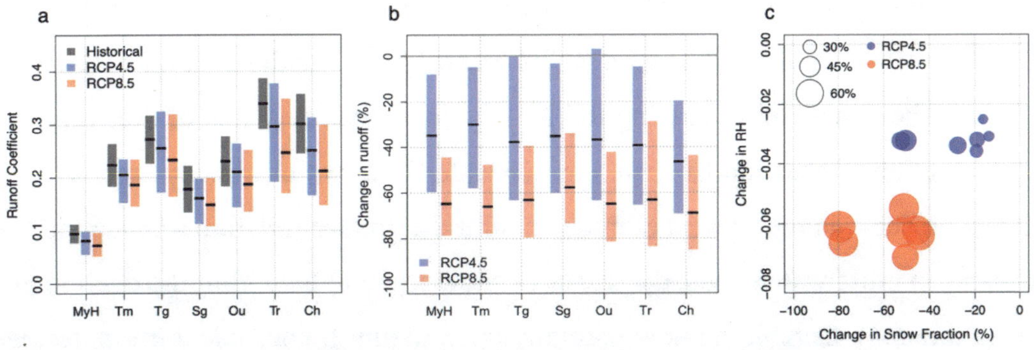

Figure 5.13: (a) Average runoff coefficients for the seven catchments in the observations (black), and projected average values in the RCP4.5 (blue) and RCP8.5 (red) scenarios. Boxes rep- resent 90% confidence intervals. (c) Projected relative changes in runoff across the seven catchments against relative change in snow fraction (x-axis) and change in catchment-wide relative humidity (y-axis), for the RCP4.5 (blue) and RCP8.5 (red) scenarios.

Catchment	M. Hassan		Tamesmate		Tillouguite		Segmine		Ouchene		Tarhat		Chacha	
	RCP4.5	RCP8.5	RCP4.5	RCP8.5	RCP4.5	RCP8.5	RCP4.5	RCP8.5	RCP4.5	RCP8.5	RCP4.5	RCP8.5	RCP4.5	RCP8.5
$\Delta T(°C)$	2.3	4.6	2.2	4.4	2.2	4.5	2.3	4.6	2.3	4.6	2.3	4.8	2.3	4.7
ΔPr (%)	-11	-53	-14	-57	-30	-57	-22	-48	-31	-57	-31	-50	-31	-55
ΔRH (%)	-3.1	-6.2	-2.5	-5.5	-3.2	-6.4	-3.6	-7.1	-3.4	-6.3	-3.2	-6.6	-3.2	-6.1
ΔSnow (%)	-3.3	-11.3	-3.9	-12.3	-6.7	-15.7	-5.7	-14.9	-6	-11.3	-5	-7.5	-2	-3.2
ΔSF (%)	-14	-46	-16	-51	-19	-44	-19	-51	-28	-52	-52	-78	-50	-80
ΔRC (%)	-14	-23	-8	-17	-6	-14	-10	-17	-9	-19	-13	-27	-15	-30
ΔRunoff (%)	-23	-64	-21	-64	-34	-63	-29	-57	-37	-65	-39	-63	-41	-68

Table 5.4: Long-term (2071-2100 minus 1976-2005) projections for the seven catchments: October-May temperature (°C), precipitation (%), relative humidity (%), absolute and relative snow fraction (%), runoff coefficient (%) and runoff (%), under RCP4.5 and RCP8.5.

Future runoff projections are characterized by consistent, steep declines in runoff coefficients of 5-17% under RCP4.5 and 15-30% under RCP8.5 (Figure 5.13-a). The impact of decreasing RH largely dominates over that of declining snow fraction. Tarhat and Chacha, the two watersheds which already receive almost no snow in the present climate, exhibit the greatest relative RC decline, whereas in other watersheds, decreases in snow fraction help limit the decline in RC (Table 5.4). Combining now precipitation and runoff coefficient estimates, we find a 20-40% decrease in runoff in the RCP4.5 experiments, compared to a 50-65% decline under RCP8.5 (Figure 5.13-b, Table 5.4). Decreases in precipitation drive most of the runoff trends, especially in the RCP8.5 scenario. Projected RC declines are about the same when using CHIRPS data for the analysis, although tend to be slightly higher, on average by 3-6% on average (not shown).

Discussion

Snowpack projections

The steep projected decline in snowpack in the High Atlas results both from a decrease in wet-season precipitation, and from warming trends which severely reduce the fraction of solid precipitation. This is particularly true at mid-elevations (2000-2500 m), very close to the zero-degree line in the current climate, where the relative decline in snow fraction is the highest (Fig. 5.7). At first order, changes in precipitation and rising temperatures each seem to account for about half of the projected SWE trends for the area as a whole. Precipitation goes down by about 25% under RCP4.5 and 40-45% under RCP8.5 (Chapter 4), while the fraction of solid precipitation decreases by 25% under RCP4.5 and 50% under RCP8.5 (from ≈19% in the historical simulations to ≈14% and ≈9% annually in the RCP experiments) (Fig. 5.7). Rising temperatures prevent snowpack build-up not only by increasing the likelihood of liquid precipitation, but also by favouring snowmelt. Snowmelt is already frequent during winter in the High Atlas, even at high elevations (T20a), and is thus likely to be even more common in a warmer world. Still, at high elevations, the relative decline in snow fraction is roughly half as small, and consequently the precipitation signal may be dominant in the SWE trends.

Inter-model uncertainty in normalized SWE projections differs across scenarios (Fig. 5.11). It is small in the RCP8.5 simulations: peak SWE declines by 85±3%. In this scenario, warming is large enough in all the models that regardless of its magnitude, snowpack is projected to disappear almost entirely. All three models agree on a ≈ 50% decrease in snow-to-precipitation fraction. This is consistent with observations that in mountain regions, above-freezing temperatures are common, even at high altitudes, where average winter temperatures are not far from freezing (López-Moreno et al., 2017). Therefore, under the RCP8.5 scenario, the projected 4°C warming (Chapter 4) would regularly bring all areas but the very highest peaks well above freezing, and prevent seasonal snowpack accumulation. By contrast, uncertainty is high in the RCP4.5 simulations, in which warming and drying are about half as small as under RCP8.5 (Table 5.4). Consequently, the reduction in the snow-to-precipitation fraction projected by the models ranges from 5 to 35%. In this scenario, the uncertainty in warming thus appears to be the main factor behind the spread in snowpack projections.

There is also uncertainty in absolute SWE values. As discussed in T20a, the historical SWE peak of 125 MCM is subject to caution; experiments with enhanced precipitation or reduced temperatures suggest it may in fact be as high as 200 MCM. Still, while inter-model spread is large in the historical experiments (≈60 MCM) and still large (≈40 MCM) in the RCP4.5 experiment, it is reduced to almost zero under RCP8.5. Again, despite the spread in temperature and precipitation projections among the three models, they all agree on the virtual disappearance of the snowpack by the end of the century under business- as-usual.

Runoff modeling and projections

Among the several environmental variables tested, runoff coefficients in the High Atlas appear primarily affected by RH and the fraction of snow in seasonal precipitation. Consistent with previous studies (Wang et al., 2016; Duan et al., 2017), the higher RH is, the larger the runoff coefficient. Higher RH in the High Atlas is indeed associated with more precipitation, higher soil moisture levels and a lower evaporative demand. Thus higher RH will lead to reduced relative sublimation losses (Fig. 8-a), but it is also linked to warmer temperatures which tend to decrease the snow-to-precipitation ratio and lead to lower relative sublimation losses as well. RH is therefore important in our model because it incorporates information on both the energy constraint (evaporative demand) and on the water availability. The snow fraction has an opposite effect: the larger it is, the lower runoff efficiency. We can understand the influence of snow fraction by noting that a higher snow fraction means more opportunity for sublimation, particularly large at high elevations, and evaporation of melted snow, consistent with our analysis of sublimation losses (Fig. 5.11). Precipitation in the area tend to occur in short and intense storms, and quickly saturate the dry soil, leading to rapid overland flow with limited opportunity for evaporation (El Khalki et al., 2018). By comparison, snowmelt is slow and leads to a more gradual surface flow with the potential for higher evaporative losses in this climate where evaporation tends to be water-limited. More winter snow may also lead to a higher likelihood of rain-on-snow episodes in spring, known to cause rapid flooding due to high runoff efficiencies (Davenport et al., 2020). Whether this occurs in the High Atlas remains to be shown, but it would nevertheless be consistent with the effect of the snow fraction on runoff efficiencies.

More snow may also mean more opportunity for infiltration and aquifer recharge (Hssaisoune et al., 2020). It should be noted that karstic areas are also quite frequent within the Oum-Er-Rbia watershed (Akdim, 2015), with important implications for infiltration, aquifer and spring regimes in our study area. Compared to surface runoff, groundwater remains a small fraction of water use in the Oum-Er-Rbia basin (<15%), and even less of available renewable water since aquifers are largely overdrawn (Hssaisoune et al., 2020). Groundwater data are quite scarce in this region; we make the choice of purely focusing on surface runoff, keeping in mind that a more complete picture of basin-wide water availability would also require taking aquifer fluxes into account. While base flow is negligible in six out of the seven sub-catchments considered here, aquifer discharge may still occur naturally further down the mountains, or artificially via direct groundwater pumping in the agricultural plains.

Results indicate a decline in runoff under future climate scenarios, due to a decrease in both runoff efficiency and wet-season precipitation. Precipitation trends play a much larger role than RC changes in runoff projections. In RCP8.5, precipitation decreases by 48-57%, compared to a 15-30% decline in RCs. Our projected runoff trends are consistent with those of El Moçayd et al. (2020) who focused on catchments part of the Sebou watershed, where snow plays a much smaller role than in our region (Marchane et al., 2015). This is not surprising given the weak impact of the declining snow fraction on runoff coefficients, compared to that of the relative humidity decline. Runoff efficiency is projected to decline mainly due to atmospheric drying (decreasing RH), which more than compensates for the diminishing snow fraction. Under RCP8.5, average relative humidity will decline by 3-6% (Chapter 4), which will lead to an average 30% RC decline across the seven sub-catchments, while the decline in snow fraction increases RCs by only 10%. Additionally, lower RH implies that sublimation rates will be higher when snow is present. However, because the snow-to-precipitation ratio will also sharply decline due to rising temperatures, the overall loss of annual precipitation by sublimation will tend to decrease by about a third (Fig. 5.8-b).

The spread in runoff projections is large (Fig. 5.13-b), due to uncertainties in both RC (Fig. 5.13-a) and precipitation projections. Despite warming and drying trends being smaller in RCP4.5, the uncertainty on RC changes is just as high as under RCP8.5 (Fig. 5.13-a), notably because snowpack projections are much more uncertain (Fig. 5.11). Combined with weaker precipitation changes, this translates into larger uncertainty in future runoff under RCP4.5 compared to RCP8.5.

Uncertainties in results and limitations of the study

Every modelling framework relies on its own assumptions, which translate into uncertainties. First, it is important to note that our pixel-by-pixel approach to bias-correction does not take spatial correlation and inter-variable dependencies into account, for which a more complex method like R2D2 (Vrac, 2018) would be required. However, in the present case, it makes little practical difference. To fit the snow model, we rely on average annual cycles of snow cover over which short-term correlations across space and between variables, beyond the seasonal cycle, have little influence. In addition, the target datasets used in the bias-correction come from different sources (TRMM, MODIS, etc.), and thus we should be very careful about relying on their correlations when correcting the data. Ideally, one could correct all variables from a single observational dataset that includes all the variables we use, but such a dataset obviously does not exist. Second, for future climate projections, we rely on a single regional climate model, and thus do not explore the full range of uncertainties related to model configuration and parametrisations. Still, the regional simulations we use here have been specifically tailored to the area, particularly the choice of driving GCMs, and validated against a range of observations (Chapter 4). The fact that the uncertainty in snowpack projections under RCP8.5 is small could be further explored by using other GCM/RCM combinations, especially ones that lead to less warming than projected in our three-member ensemble. As to the uncertainty in precipitation trends, it is difficult to reconcile. As shown in Chapters 2 and 3, the magnitude of future wet-season precipitation in Northwestern Africa is mainly determined by that of changes in Mediterranean atmospheric circulation. Spread in dynamical trends is difficult to reduce for this region. A GCM-selection approach based on storylines could be relevant to determine plausible scenarios (Shepherd, 2019). In addition, drier soils together with enhanced absorption of solar energy where snowpack disappears will also lead to enhanced warming locally, driving yet further snowpack melt. We do not explicitly take this into account in our model. In particular, the snow albedo effect is largely absent from the MRCM simulations due to their 12 km resolution, which is still too coarse to represent the complex topography. For areas at the highest elevations (near 4000

m) which will still likely be below freezing in future winters, melt may remain largely unchanged in the middle of winter; however, a drier atmosphere will still be associated with reduced precipitation and increased sublimation losses, which will play a critical role in reducing the snowpack.

The advantage of using a statistical regression model to investigate the sensitivity of runoff to climate variables and its changes is its simplicity and interpretability. It allows for instance to highlight the influence of the snow fraction in the inter-annual variability of RCs. While hydrological models are certainly more comprehensive, they also rely on many parameters which leaves them vulnerable to overfitting and thus to even more uncertainties. In addition, they suffer from the same drawback as our simple model, which is that there is no guarantee that the relationships obtained in the current climate are directly transferable to the future climate. Still, the large inter-annual variability in climate variables in the present climate helps make the model more robust to large changes in average covariate values. Furthermore, the explanatory variables used in the regression are physical variables which we understand how mechanistically relate to runoff. The difference with a hydrological model is in the different form of mathematical relations, however the physical relationship between the variables is somewhat respected. Furthermore, RH and snow fraction only explain 30% of inter-annual variability in RCs. The rest of the variability may be still explained by climatic variables, which we do not include in our model. Our seasonal approach to defining predictor variables is also quite restrictive; it does not take into account the weekly variability (in RH, temperature, precipitation, etc.) which may affect runoff efficiency. For instance, the details of the daily distribution of precipitation are important to estimate runoff: the more concentrated in time precipitation is, the less opportunity for infiltration and evaporation, and therefore the higher the runoff efficiency.

Finally, we focus on climate-driven inter-annual variability in RCs; thus, we do not explain differences in average RCs across catchments, nor can we say anything about long-term variability in RCs forced by non-climatic parameters, like land use change. This last aspect may not be critical in the case of the High Atlas since most of the area under study is uncultivated, naturally lacking tree cover and sparsely populated. It has thus not experienced very significant land use changes in the last few decades. However, climate-change driven trends in vegetation cover may still affect runoff efficiency in the region. It is also unclear to what extent enhanced groundwater pumping in the Oum-Er-Rbia watershed since the 1980s may have modified the natural water balance and, indirectly, RCs in mountain catchments.

Conclusions

Based on the robust understanding of its snow water balance in the current climate, we quantified in this chapter the response of the High Atlas snowpack to climate change using high-resolution downscaled climate projections. We find that the High Atlas snowpack will significantly decline, even in the RCP4.5 scenario which involves substantial mitigation of emissions. By the end of the century, snow may become a rarity below 2000m. Maximum snow cover extent is projected to go from 17% of the study area down to 9% under RCP4.5 and even 4% under RCP8.5 (Fig. 5.9-f). In parallel, snowpack water equivalent could decline by 80%, even at the highest elevations (>3500 m). Snowpack decline is evidently connected to regional warming trends, but also affected by a projected 40-60% decrease in wet-season precipitation in Northwestern Africa.

The analysis of runoff coefficients for seven mountain catchments showed that a third of their inter-annual variability could be explained by large-scale meteorological factors like snow fraction of precipitation and relative humidity. Interestingly, in this region, a larger snow fraction leads to lower runoff coefficients. While the reverse is generally thought to be true at higher latitudes (Berghuijs et al., 2014), this finding is consistent with other analyses in warm, semi-arid regions that receive substantial amounts of snow during winter (Davenport et al., 2020). Warmer conditions tend to enhance runoff efficiencies by reducing snowpack, thus limiting sublimation losses and the slow melting of snow, propitious to evaporation, and by increasing the likelihood of rain-on-snow events that tend to cause high runoff efficiencies. While decreasing snowfall will partly compensate for the projected atmospheric drying over the region, runoff coefficients will tend to decline by 5-30% depending on catchment and scenario. Combined with precipitation trends, basin-wide runoff could be reduced by 60% in the worst case.

The robust physical understanding behind large-scale projections for the region (see Chapter 2) increases the likelihood that the dire projections detailed above will be realized provided greenhouse gas emissions are not brought under control. This would deal a severe blow to the region, jeopardizing its agriculture-based economy and the livelihood of millions of smallholder farmers. Agriculture, which accounts for 90% of current water use, will have no choice but to adapt. A 40-60% precipitation decline would make rainfed agriculture infeasible. At the same time, availability of water for irrigation will also decline sharply. A change of cropping patterns, like a transition to tree crops with less water demand and higher economic value like olives, will likely be unavoidable for Morocco to adapt to this future reality.

A version of this chapter first appeared as:
Tuel, A., El Moçayd, N., Hasnaoui, M. and Eltahir, E., 2022. Future projections of High Atlas snowpack and runoff under climate change. Hydrology and Earth System Sciences, 26(3), pp.571-588.

Chapter 6: Projections of Intensifying Drought over the Eastern Mediterranean

Suchul Kang
Alexandre Tuel
Elfatih A. B. Eltahir

The Eastern Mediterranean region (defined here as 32.1°E to 39.1°E, and 31°N to 36.3°N; Figure 6.1.) is a hotspot for political conflict due to a combination of social, historical and environmental factors, among which population growth, dry climate, and water scarcity (Gleick 1994; Allan and Allan 2002; Bolle 2003). The volume of renewable available water per person is among the lowest compared to any other area (FAO 2020). During the last 30 years, the Eastern Mediterranean has experienced significant and widespread drought conditions (Cook et al. 2016; Hoerling et al. 2012; Scheffran and Battaglini 2011; Kelley et al. 2015; Chenoweth et al. 2011). The recent drought in Syria may have contributed to the deteriorating economic and social conditions that fueled conflicts in this unstable region. Although it remains unclear whether this drought is in any way linked to climate change, both Regional Climate Models (RCMs) and Global Climate Models (GCMs) robustly project a shift towards enhanced dry conditions for this

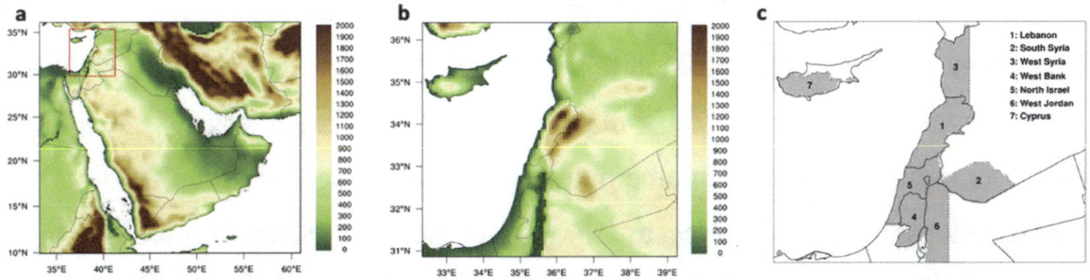

Figure 6.1: Model domain and topography used for the (a) MRCM/25km and (b) MRCM/7km. The red box in (a) denotes the localization of the nested 7km Eastern Mediterranean domain. (c) The seven sub-regions used in this study.

region (Giorgi and Lionello 2008; Chenoweth et al. 2011; Önol et al. 2014; Zappa et al. 2015; Bucchignani et al. 2018; Ozturk et al. 2018; Chapter 2). Given its current level of water stress, any further reduction in water availability will severely complicate the situation in the Eastern Mediterranean. Therefore, it is important to understand why and to what extent the future climate may result in drier conditions in the region. The answers to these questions will be crucial to devise adaptation measures and policies to mitigate the impact of increasing water scarcity. Effective adaptation for the region will require accurate projections of temperature and precipitation totals, and their spatial distribution.

Several studies have developed regional climate projections for the Eastern Mediterranean by relying on dynamical downscaling with RCMs, including within coordinated frameworks (e.g., Mediterranean (MED) and Middle East North Africa (MENA) Coordinated Regional Climate Downscaling Experiment (CORDEX), Ruti et al. 2016; Zittis et al. 2014). Early analyses by Önol et al. (2014) highlighted a pronounced precipitation decrease along the Eastern Mediterranean coastline under future emissions scenarios, but they did not test the ability of their RCMs in simulating the current climate of the region at the selected 27km resolution. Bucchignani et al. (2018) analyzed climate projections from COSMO-CLM at 0.22° and 0.44° resolution over the CORDEX-MENA domain under RCP 4.5. Both resolutions led to similar large temperature increases, while the decline in precipitation was largely dependent on the horizontal resolution and noticeably lower over the Eastern Mediterranean at high resolution. Ozturk et al. (2018) also projected warmer and drier conditions over the same domain using RegCM4.4 at 50 km resolution under RCP4.5 and RCP8.5. Still, the resolutions employed in these studies, while admittedly better than those of GCMs, remain relatively coarse, particularly for the Eastern Mediterranean, characterized by sharp topography gradients and narrow coastal mountain ranges that largely drive the spatial variability in precipitation.

For a region as complex as the Eastern Mediterranean, high resolution is key to accurately simulate the spatial and temporal variability of precipitation. A significant body of studies have demonstrated the positive impact of higher resolution on precipitation simulations over complex orography within RCMs (Cardoso et al. 2013; Chan et al. 2013; Fosser el al. 2015; Kendon et al. 2017; Liu et al. 2017; Saeed et al. 2017; Brisson et al. 2016; Zittis et al. 2017). In particular, Zittis et al. (2017) demonstrated the potential of very high horizontal resolution (12km, 4km, 1km) in improving the simulation of daily extreme precipitation over the Eastern Mediterranean with the WRF model. Therefore, any attempt at developing climate change projections for this region should seek resolutions below 10km. Hochman et al (2018) generated future climate projections at 8km over Israel using a double nesting method with COSMO-CLM nested into a 25km-resolution CORDEX-MENA simulation. Still, their study only analyzed a small part of the Eastern Mediterranean, and only under the RCP4.5 scenario. In addition, the physical mechanisms responsible for projected precipitation changes have not been discussed in detail, either at local or at regional scales.

In this study, we seek to project and improve our understanding of future changes in surface climate variables under global climate change over the Eastern Mediterranean using the Massachusetts Institute of Technology (MIT) regional climate model (MRCM). Regional climate change projections under RCP8.5 and RCP4.5 are carried out using a nest-down approach, with two resolutions: 25km and 7km. For ensemble regional projections, we generate three MRCM projections driven by three different GCMs that are carefully selected through a rigorous evaluation of their reference simulations (see section on AOGCM selection below). Future drought conditions ("drought" being defined as periods when annual precipitation is lower than its long-term historical average) are discussed for the two GHG concentrations scenarios, with a specific focus given to the strong wintertime drying projected for this region. Finally, we present and discuss the physical mechanisms responsible for the projected dry conditions in this region. Two important aspects of this study are novel relative to previous studies. The first aspect is the careful analysis of the AOGCMs selected for use in prescribing the boundary conditions in order to make sure that these conditions are simulated by models that have been tested and verified in their ability to simulate adequately climate conditions in this region. The second is the theoretical contribution emerging from contrasting the high-resolution simulations performed in this study with newly proposed physical mechanisms responsible for the reduction in precipitation projected by these simulations.

Model description and Experimental Design

The MIT Regional Climate Model

The RCM used in this study is the Massachusetts Institute of Technology Regional Climate Model (MRCM; Im et al. 2014), which is based on the Abdus Salam International Centre for Theoretical Physics Regional Climate Model Version 3 (RegCM3; Pal et al. 2007). MRCM includes several significant improvements (Winter el al. 2009; Im et al. 2014; Pal and Eltahir 2016), such as a new land surface scheme, and accurately modified soil albedo and emissivity over desert and semi-desert. In particular, Pal and Eltahir (2016) applied MRCM to Southwest Asia and demonstrated that it performed reasonably in reproducing the regional climate variables over this region, where land types are similar to those of the Eastern Mediterranean.

Simulation domain and experimental design

Our simulation domain for this study is the Eastern Mediterranean, here defined as extending from 32.1°E to 39.1°E and 31°N to 36.3°N (Figure 6.1). The climate of the Eastern Mediterranean is generally arid with hot and dry summers, and mild winters (Bolle 2003). Precipitation occurs mostly during winter over coastal regions. Inland areas receive very little precipitation year-round. The spatial distribution of precipitation is dictated by the location of the storm tracks characterizing weather systems advancing eastward from the Mediterranean and their interactions with the mountains and hills scattered along the coast from Western Syria to the West Bank of the Jordan river (Figure 6.2d). As a result, observed precipitation exhibits a sharp gradient of about one millimeter per kilometer, decreasing from about one meter per year next to the coast to about one

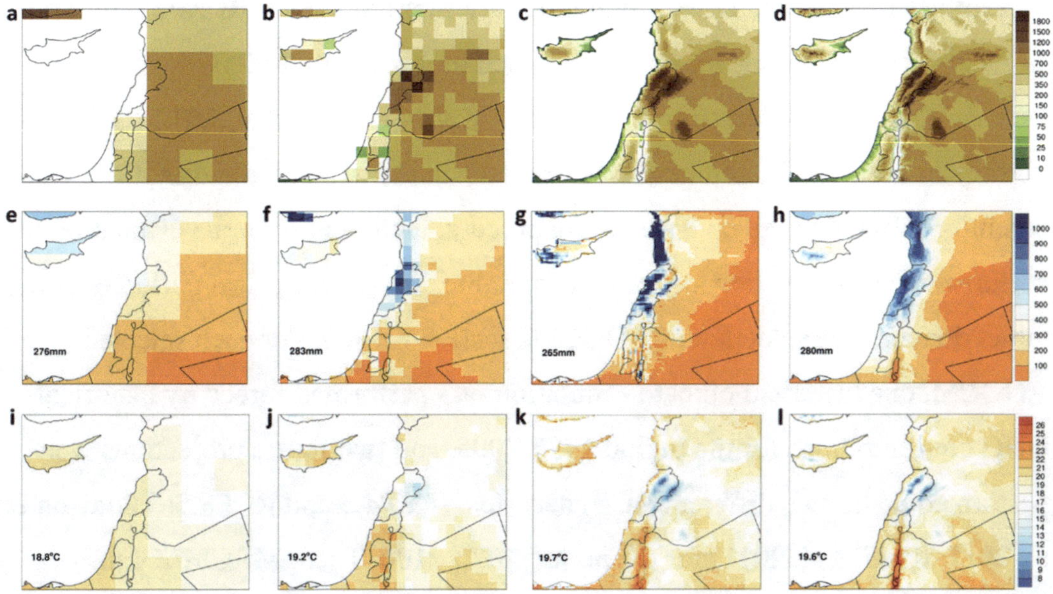

Figure 6.2: (a)-(d): Eastern Mediterranean topography from (a) GTOPO30 100km, (b) MRCM/25km, (c) MRCM/7km, and (d) GTOPO30 1km. (e)-(h): Mean annual total precipitation from (e) GCMs interpolated onto a 100km grid, (f) MRCM/25km, (g) MRCM/7km, and (h) TerraClimate (5km). (i)-(l): Mean annual surface temperature from (i) GCMs interpolated to a 100km grid, (j) MRCM/25km, (k) MRCM/7km, and (l) TerraClimate (5km). Mean annual total precipitation and annual surface temperature are calculated using data from a 30-year period (1976-2005). GCMs, MRCM/25km, and MRCM/7km represent an ensemble mean based on the arithmetic average of the three different simulations. Area-averaged value over land is included in the bottom left corner of (e)-(l).

hundred millimeters per year in the interior deserts (Figure 6.2h). This sharp gradient of precipitation distribution is the most important feature characterizing the climate of this region, and constrains water availability for its people. As expected, the spatial distribution of temperature closely follows topography in this region (Figure 6.2-l).

We carry out regional climate change projections over our domain with MRCM by adopting a nest-down approach at two resolutions: 25km and 7km (Figure 6.1a-b). We use the 25km-resolution regional climate simulations performed by Pal and Eltahir (2016) over Southwest Asia, which includes the

Eastern Mediterranean. These simulations are referred to as MRCM/25km. The surface and lateral boundary conditions used to force MRCM/25km were taken from 3 AOGCMs participating in CMIP5: CCSM4 (Meehl et al. 2012), MPI-ESM (Giorgetta et al. 2013), and NorESM (Bentsen et al 2013). These AOGCMs were carefully selected based on a screening process, including rigorous evaluation of performance in simulating key climate variables for the historical period over the target domain (see selection of AOGCMs section below). For each selected AOGCM, one historical climate simulation was performed forced by historical GHG concentrations for the period 1975-2005, and two future projections were performed under two different RCP scenarios, (RCP4.5 and RCP8.5; Thomson et al. 2011; Riahi et al. 2011) for the period 2070-2100. Then, MRCM/25km simulations are themselves used as lateral boundary conditions for a nested 7km simulation over the smaller Eastern Mediterranean domain (MRCM/7km) shown in Figure 6.1b, over the period 1975-2005 for historical, and 2070-2100 for future, respectively. No bias correction of output variables is performed.

AOGCMs selection

The three AOGCMs used in MRCM/25km as lateral boundary conditions are selected among the 40 AOGCMs in CMIP5 archive by applying the following criteria (Pal and Eltahir, 2016): 1) The AOGCM provides representation of the Red Sea and Persian Gulf by use of an ocean model of adequate resolution. They selected 7 AOGCMs out of the 40 by requiring an oceanic horizontal resolution of 1.11° or higher, capable of simulating some of the key processes in the Red Sea and Persian Gulf. 2) Surface variables (surface temperature (T), wet-bulb temperature (TW), and relative humidity (RH)) from 7 AOGCMs are objectively

Model	Atmosphere (latitude x longitude)	Ocean (latitude x longitude)	Main reference
Community Climate System Model Version 4 (CCSM4)	0.9° x 1.25°	1.11° x 0.27° ~0.54°	Gent et al. 2011
Max-Planck-Institute Earth System Model Medium resolution (MPI-ESM-MR)	T63(~1.875°)	0.4° x 0.4°	Giorgetta et al. 2013
Norwegian community Earth System Model (NorESM)	1.875° x 2.5°	1° x 1°	Bentsen et al. 2013

Table 6.1: Descriptions of Global Climate Models used in this study.

analyzed and compared to the ERA-Interim and CRU datasets. The normalized root-mean square errors (RMSE) of these surface are averaged and compared separately over the Red Sea, Persian Gulf, and Arabian Peninsula. Based on the total sum of RMSE for each variable and region, 3 AOGCMs are selected: Community Climate System Model version 4 (CCSM4), Max-Planck-Institute Earth System Model (MPI-ESM) and Norwegian community Earth System Model (NorESM). More detailed information about the GCMs is presented in Table 6.1.

Results

Evaluation of MRCM/7km for the reference period

To evaluate the performance of MRCM/25km and MRCM/7km, the simulated surface temperature and precipitation were compared with the TerraClimate dataset (Abatzoglou et al. 2018), focusing on the seasonal means and annual cycles over 7 sub-regions (Figure 6.1c). TerraClimate consists of global high-resolution (~5km) monthly climate and climatic water balance data over terrestrial surfaces from 1958 to 2017. For a more in-depth analysis of precipitation, in-situ observations at six stations from ECA&D (European Climate Assessment & Dataset, Haylock et al. 2008; Klein Tank et al. 2002), are also used (Figure 6.3). The six stations are selected based on their location (three stations at low elevation and three stations at high elevation) and data availability (we require less than 10% of missing values).

A good test for models used in climate studies over this region is to assess their ability to reproduce observed spatial distributions of temperature and precipitation, including the west-east gradient of precipitation characteristic of the Eastern Mediterranean. As shown in Figure 6.2-a, the topography of the region in coarse-resolution (100-200km) models bears little resemblance to reality. Consequently, global models unsurprisingly fail to reproduce the observed patterns of temperature and precipitation. The topography in MRCM/25km is shown in Figure 6.2-b. Although the simulated patterns of temperature and precipitation improve compared to GCM simulations (Figure 6.2-f,j), there remains significant differences with observations. Precipitation is largely underestimated along the coast, and overestimated in the interior deserts. In

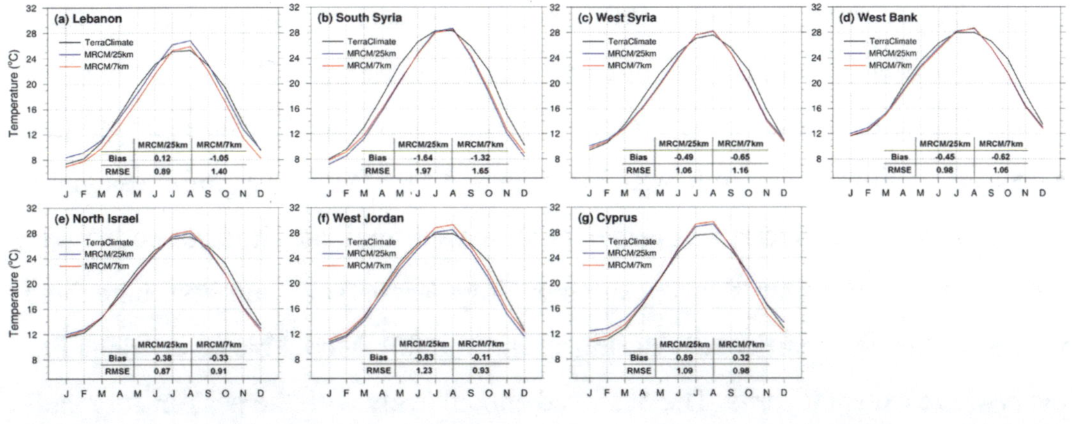

Figure 6.3: Annual cycle of surface temperature from TerraClimate (black), MRCM/25km (blue) and MRCM/7km (red) over the seven sub-regions. Table in each plot includes Bias and RMSE calculated between MRCM (25km and 7km) and TerraClimate data. Monthly mean surface temperatures are calculated using data from a 30-year period (1976-2005).

MRCM/7km, however, the distribution of precipitation and temperature exhibit significant improvements, and match observations much more closely (Figure 6.2-g,k). In particular, the observed west-east gradient of precipitation is reasonably well reproduced, though some discrepancies can still be detected along the coast.

Similarly, patterns of annual-mean temperature in MRCM/7km are much closer to TerraClimate than those shown in GCMs or even MRCM/25km (Figure 6.2-i,j,k,l). GCMs show only a general warming trend as latitude decreases, naturally missing the local temperature extrema associated with the topography (Lebanese mountain ranges and the Jourdain Valley). MRCM/25km already brings noticeable improvement in the temperature pattern, but still lacks much of the detail associated with topography. By contrast, the MRCM/7km simulation shows a remarkable match with TerraClimate over much of the domain. In particular, the complex temperature pattern over Lebanon and Western Syria, and between Israel and Jordan are correctly reproduced, even though desert regions to the east still appear to suffer from a general cold bias (Pal and Eltahir 2016).

Figure 6.3 presents the annual cycle of area-averaged surface temperature across all 7 sub-regions. Both MRCM/25km and MRCM/7km capture reasonably well the seasonal temperature cycle over all the sub-regions, although both tend to overestimate temperatures during summer (July-August) and underestimate them the rest of the year, particularly during spring and fall. Model biases are negative everywhere except Cyprus. The observed annual cycle with the peak in July and August is well captured in amplitude and timing in both MRCM/25km and MRCM/7km, but the annual cycle produced by MRCM/7km is closer to the observed one with slightly less bias and RMSE in 3 sub-regions (South Syria, West Jordan and Cyprus). Overall, MRCM/7km tends to exhibit lower temperatures than MRCM/25km, but the difference is small (<0.5°C), except except for Lebanon, West Jordan and Cyprus. This difference likely comes from a better simulation of precipitation amounts over these regions, which leads to enhanced latent heat fluxes and an overall decrease in mean temperatures, especially during the wet season.

Regarding the annual cycle of precipitation (Figure 6.4), results are mixed. TerraClimate precipitation peaks in January in most of the sub-regions, excepted in West Syria and Cyprus where the peak occurs in December. In general, MRCM/25km and MRCM/7km are able to capture some of the differences in the annual cycle from region to region. For example, Syria is characterized by substantial difference in precipitation pattern between its southern and western parts. Still, apart from Western Syria, where MRCM/7km brings a significant

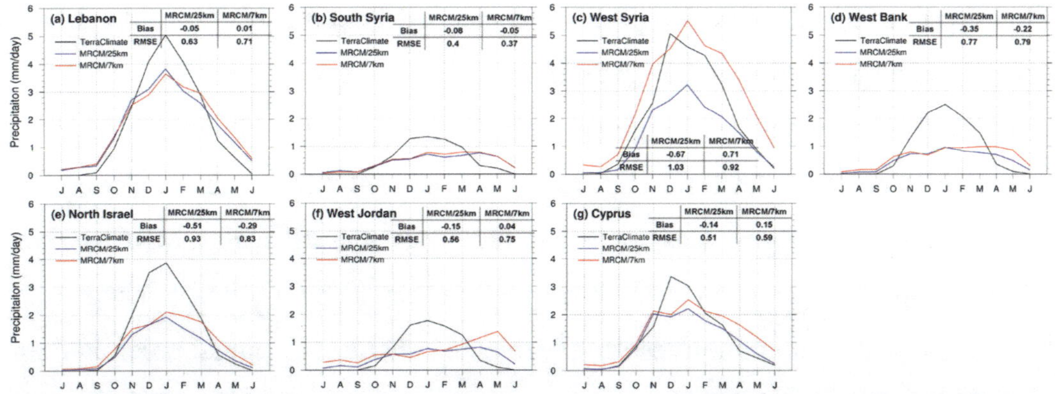

Figure 6.4: Annual cycle of precipitation from TerraClimate (black), MRCM/25km (blue) and MRCM/7km (red) over the seven sub-regions. Table in each plot includes Bias and RMSE calculated between MRCM (25km and 7km) and TerraClimate data. Monthly mean precipitations are calculated using data from a 30-year period (1976-2005).

improvement to the simulation of precipitation (Figure 6.5-c), we do not find substantial differences between the 25km and 7km simulations for most sub-regions and significant discrepancies between simulated and observed precipitation persist even for the high-resolution simulations, especially for northern Israel, the West Bank and western Jordan where the simulations exhibit an evident dry bias essentially independent of the resolution. Because we rely on GCM output as boundary conditions, part of that bias undoubtably comes from biases in the GCMs themselves. In particular, the dry biases are most pronounced to the south of our domain, where GCMs possibly misrepresent the direction and magnitude of moisture transport. Additionally, despite their advantages, gridded observational data unavoidably contain some errors due to the interpolation method in mountainous and coastal regions, where stations are typically scarcer and the terrain more complex. The comparison of MRCM simulated precipitation with observations at station locations shows that while annual cycles are generally well-reproduced, as in the case of the sub-regions dry biases persist even in the

7km experiments and resolution does not appear to affect this bias considerably.

Future projections of intensifying drought

Still, MRCM/7km simulates a regional-scale precipitation (and temperature) pattern much closer to the TerraClimate dataset than MRCM/25km, let alone GCMs (Figure 6.2-e,f,g,h). Despite its shortcomings at small scales, MRCM/7km can offer important perspectives on precipitation trends under climate change at the regional scale. Simulation results of projected changes in annual precipitation are shown on Figure 6.5 for a business-as-usual scenario (RCP8.5). All three

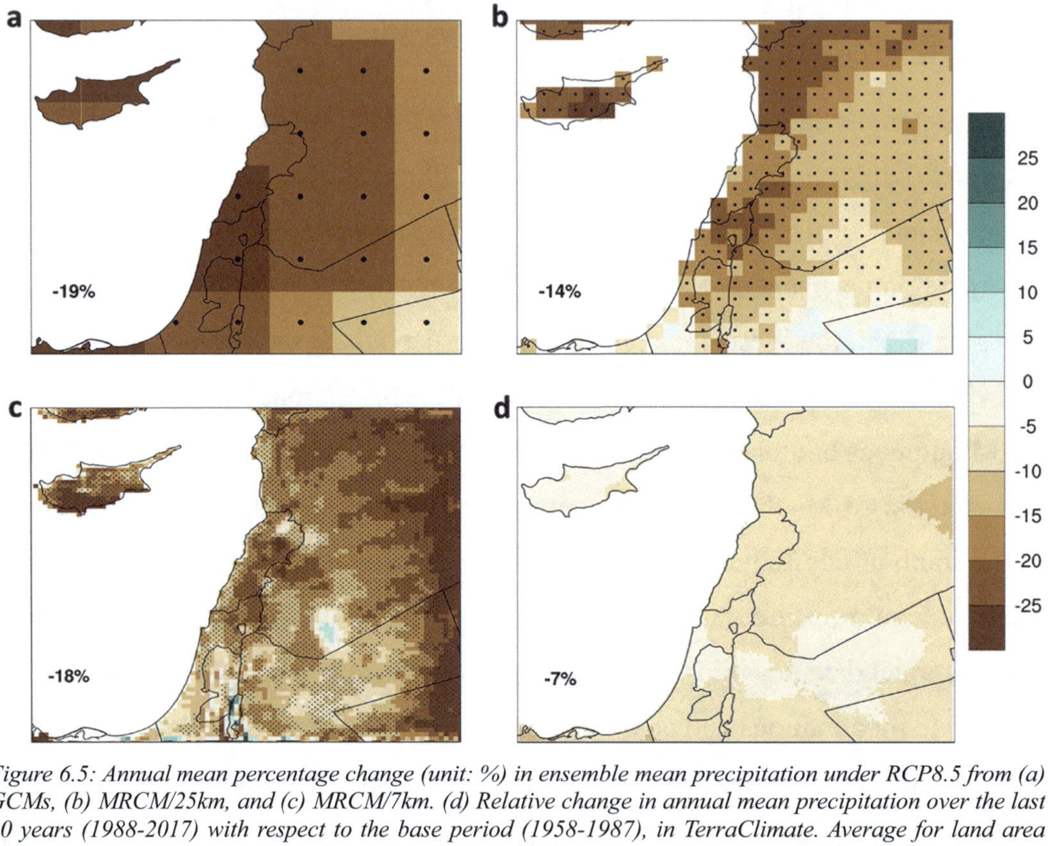

Figure 6.5: Annual mean percentage change (unit: %) in ensemble mean precipitation under RCP8.5 from (a) GCMs, (b) MRCM/25km, and (c) MRCM/7km. (d) Relative change in annual mean precipitation over the last 30 years (1988-2017) with respect to the base period (1958-1987), in TerraClimate. Average for land area excluding buffer zone is indicated in each plot. In (a)-(c), dots indicate agreement by all three ensemble members on the sign of the change.

members unequivocally project more intense future drought conditions over the domain. Still, important differences can be noted. While both GCMs and MRCM/25km indicate larger reductions in annual precipitation along the coasts, with the intensity of the change decreasing as one moves inland (Figure 6.5-a,b), the high resolution (MRCM/7 km) simulations show relatively uniform drought conditions, with drought intensity that does not particularly decrease with longitude (Figure 6.5-c). However, right at the coast, from Syria to Israel, MRCM/7km points to slightly smaller reductions precipitation than GCMs and MRCM/25km. This may be due to the latter two largely underestimating coastal precipitation to begin with. With its improved representation of topography, one also expects the nature of precipitation in MRCM/7km to be more weighted toward orographic precipitation than large-scale, as in GCMs, which suggests that orographic precipitation along the coasts may be less sensitive to climate change than large-scale precipitation. For Cyprus, however, all simulations project severe declines in precipitation, though MRCM/7km points to a larger relative drying on the southern half of the island. Interestingly, the pattern of precipitation change in MRCM/7km is roughly similar to that of precipitation anomalies in TerraClimate for the region over the last 30 years (Figure 6.5-d): an overall drying, but with relatively reduced intensity in a region extending from northern Israel to north-eastern Jordan. Still, the scarcity of stations in this region should make us cautious about drawing conclusions, especially when it comes to patterns of change that is typically heavily influenced by the location of available stations. In addition, the TerraClimate data is influenced by decadal variability in precipitation that does not align with model simulations.

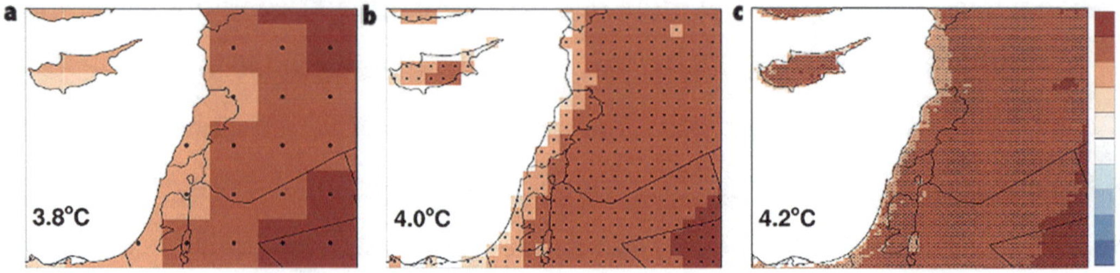

Figure 6.6: Annual mean change (unit: K) in ensemble mean temperature under RCP8.5 from (a) GCMs, (b) MRCM/25km, and (c) MRCM/7km. Average for land area excluding buffer zone is indicated in each plot. Dots indicate agreement by all three ensemble members on the sign of the change.

The projected variation in intensity of the drought among the different sub-regions is less certain than the projections about the overall regional drought and the associated spatial pattern (Figure 6.5). This uncertainty stems from the fact that the skill of the model in simulating precipitation processes at local scales is in general lower than the corresponding skill for regional scales. The projections for the mitigation scenario (RCP4.5) are similar to those of the business-as-usual scenario but with less intensity, indicating that future water availability in this region would benefit from serious mitigation efforts at the global scale.

The projected changes in annual temperature are presented in Figure 6.6, from the GCM simulations as well as the RCM simulations at the two resolutions. Since the regional average change increases slightly with resolution, the overall regional pattern of the projected change remains uniform and somewhat insensitive to resolution.

Physical mechanisms

The projected decline in precipitation over the domain, robust across all models and scenarios, follows from anomalous northeasterly flow driven by enhanced surface pressure over the central Mediterranean (Figure 6.7). The latter has itself been a consistent feature of climate projections over the area (Giorgi and Lionello 2008). The development of an anomalous Mediterranean ridge during winter and early spring results from the combination of large-scale changes in the structure of stationary planetary waves, and of the reduced warming of the Mediterranean Sea with respect to the surrounding land, as land areas are projected to warm more than oceans (see Chapter 2). The anomalous low-level northeasterly circulation has two major consequences for the region. First, its easterly component acts to weaken the background westerly flow that guides storms into the region, limiting the supply of moisture from the Mediterranean Sea. Second, because of its orientation, the anomalous wind blows downward along the steep isentropic surfaces, which leads to subsidence and enhances regional atmospheric stability and further reducing storm activity (Figure 6.7-e). In addition, the reduced warming of the Mediterranean Sea relative to land areas acts to limit the increase in evaporation and constrain moisture availability for precipitation in downstream regions (Byrne and O'Gorman 2015). Increases in static stability over the sea may also act to further limit moisture transport towards continents.

Because of the coarse resolution of GCMs, orographic precipitation is negligible compared to large-scale, cyclonic precipitation in the CMIP5 models (Figure 6.2-e). As a consequence, the increase in atmospheric stability due to northeasterly flow down the isentropes and regional forced descent plays the main role behind the reduction in precipitation in GCM projections (see Chapter 2). Enhanced subsidence can clearly be seen over the northern part of our domain in the GCMs (Figure 6.7-c). In the high-resolution MRCM simulations,

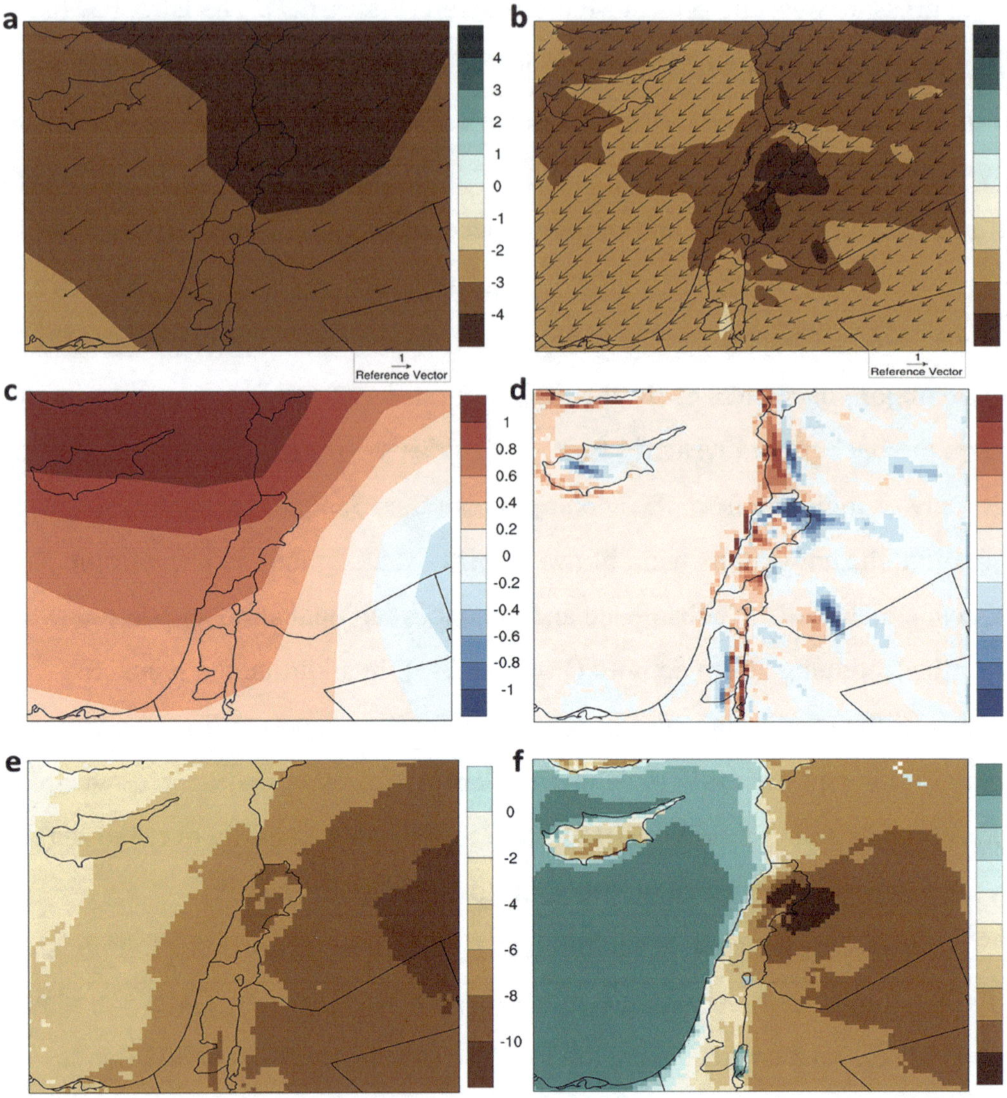

Figure 6.7(a-b): Projected changes under RCP8.5 in DJF 850 hPa wind (arrows, unit: m/s) and relative humidity (shading, unit: %) in (a) the three GCMs and (b) MRCM/7km. (c-d) Same as (a-b), but for 850 hPa vertical motion (unit: hPa/h). (e) Projected relative change (unit: %) under RCP8.5 in DJF 850 hPa 15-day filtered geopotential variance from MRCM/7km. (f) Projected change (unit: g/kg) under RCP8.5 in DJF 2-meter specific humidity from MRCM/7km.

area-average vertical motion is dominated by orographic forcing, and consequently the large-scale subsidence increase in GCM projections is considerably weakened, leaving only a small positive signal over the sea part of the domain (Figure 6.7-d). Large-scale storm activity, which is mostly set by the boundary conditions, is still reduced, particularly in the southeastern half of the domain (Figure 6.7-e). Subsidence near the top of the boundary layer is substantially enhanced downwind (with respect to the anomalous wind) of the orography, however, with corresponding uplift upwind. In addition to the weakening of the westward, precipitation-bringing flow, this appears to drive the decrease in precipitation in coastal areas and over the Mount Lebanon and Anti-Lebanon ranges. The same mechanism is at play over Cyprus and the smaller mountains of southern Syria (Jabal-Al-Druze), where mean precipitation is more moderate (Figure 6.2-h). In accordance with the above discussion, the projections exhibit steep declines in relative humidity across much of the region; local "hotspots" exist on the eastern slopes of mountain ranges that lie downwind of the prevailing westerlies (Fig 7-b). There, orographic precipitation on the peaks to the west prevents specific humidity from increasing as much as upwind (Figure 6.7-h).

Summary and conclusions

We find that the higher resolution, in general, improves the simulation of precipitation and temperature patterns over the Eastern Mediterranean due to better representation of the complex topographic features, although significant discrepancies persist. Compared to GCMs and the coarser 25km simulation, MRCM/7km can simulate much more accurately the regional distribution of precipitation and temperature over this mountainous region, especially its sharp east-west precipitation gradient. Still, the increase in resolution does not seem to reduce bias in precipitation at sub-regional scales, particularly when compared against available station data. An important source of uncertainty is the physical representations in the high-resolution model. In our simulations, we assumed that representations of physical processes in MRCM such as formation of clouds and precipitation and moist convection are valid for a range of scales down to about 50 km^2. This assumption needs to be examined carefully although probably justified here by noting the limited role that moist convection plays in precipitation formation over this region in winter. Another major source of uncertainty stems from the limited skill of the global and regional models in simulating the climate in and around this region. Here, we address some of this limitation by careful selection of global models and the use of only those models with relatively better skill in comparison to observations. GCMs remain however imperfect; and while regional models may perform satisfactorily when forced with reanalysis data, they inherit GCM biases when used for climate change projections. While convection-permitting models have been run successfully to downscale the current climate over this region, it remains to be seen whether they would be as successful when used under climate change conditions. Future research should explore cloud-permitting models with higher resolutions down to a few kilometers, and quantify biases due to the GCM forcing data.

Projections of even drier future conditions in the Eastern Mediterranean, a region that already suffers from severe water scarcity, should have significant implications to national policies regarding mitigation and adaptation to climate change. The experience of the recent conflict in this region which was associated with a prolonged drought and severe ramifications including massive human migration suggests that impacts of future climate change on water availability in this region may exacerbate the dry conditions conducive for further conflict in the region. National policies supporting global mitigation efforts as well as policies regarding sustainable patterns of population growth may offer some options for pursuing initiatives that would reduce future ecological stresses in this important region. Better preparedness for climate change adaptation would require efficient management of water demand in the future by adopting higher standards for water use efficiency in agriculture, increasing agricultural productivity and yield through better technology, and importation of virtual water.

Chapter 7: Mechanisms of European summer drying under climate change

Alexandre Tuel
Elfatih A. B. Eltahir

A large majority of climate models have long projected drier summers over much of Southern and Western Europe under future climate scenarios (Rowell and Jones 2006; Seager et al. 2014; Brogli et al. 2019; Grillakis 2019), the result of intense surface warming and a widespread decrease in precipitation, most intense between 40° and 50°N (Figure 7.1c). Such trends threaten to put significant stress on human health, water resources and agriculture (Samaniego et al. 2018), particularly in the regions of Southern Europe where precipitation will also decline during winter (Figure 7.1a). While drying trends over Europe occur in the bigger context of widespread subtropical drying, they stand out by their magnitude, much larger than over any other region (Figure 7.1d). Projections of summer drying in Europe have thus attracted significant attention in the literature, notably in the context of the recent record-breaking heatwaves and droughts that the continent experienced (Stott et al. 2004; Fischer et al. 2007; Bador et al. 2016;

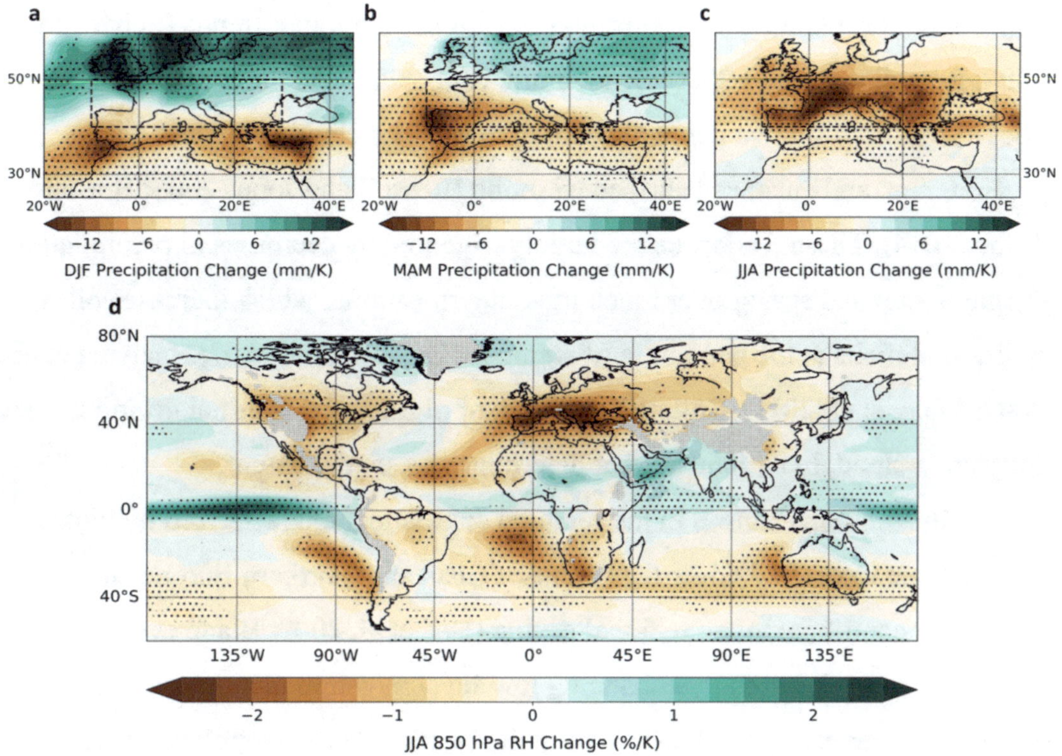

Figure 7.1: CMIP5 multi-model mean change under RCP8.5 (2071-2100 minus 1976-2005) in (a-c) European (a) DJF, (b) MAM and (c) JJA precipitation; and in (d) global JJA 850 hPa relative humidity. Projections are normalized by each model's global-mean projected temperature change. Stippling indicates agreement by at least 80% of models on the sign of the change. Grey hatching in (d) covers areas where the 850 hPa level is below the surface.

Kew et al. 2019; Vautard et al. 2020). It is now understood that positive feedbacks between a decline in soil moisture, precipitation and surface temperature play a major role in explaining the robustness of European summer trends (Schär et al. 2004; Rowell and Jones 2006; Seneviratne et al. 2006). As the surface warms, so does evapo-transpirative demand, leading to reduced cloud cover and precipitation, and enhanced surface warming (Rowell and Jones 2006; Boé and Terray 2014). These feedbacks are largely triggered by decreases in precipitation during winter and spring over much of southern Europe, which increase soil moisture deficits at the beginning of summer (Vautard et al. 2007; Haarsma et al. 2009; Wang et al. 2011). Still, the mechanism triggering this initial drying is not fully understood, in particular what determines its spatial extent.

In addition, changes in the regional circulation, characterized by higher sea-level pressure (SLP) over the Atlantic Ocean and Northern Europe, are thought to contribute to the regional drying (Boé et al. 2009; Bladé et 47 al. 2012; Belleflamme et al. 2015) by enhancing low-level moisture divergence over Europe (Seager et al. 2014) and shifting the North Atlantic jet poleward (Haarsma et al. 2015). The contribution of circulation trends appears limited to Northwestern Europe (Rowell and Jones 2006; Belleflamme et al. 2015), where it may account for more than half of the precipitation decline and much of inter-model spread (Boé et al. 2009; Haarsma et al. 2015), though other results suggested it may extend to most of the continent (Seager et al. 2014). As such, the relative importance of these circulation trends in explaining the projected drying along different latitudinal belts across Europe is not fully understood and deserves further study. The projected SLP increase in the North Atlantic results from reduced rates of ocean surface warming in that region, linked to a weakening of the Atlantic Meridional Overturning Circulation (AMOC) (Haarsma et al. 2015).

Large-scale continent-ocean warming contrasts may also further depress land relative humidity (RH) through lesser rates of increase in evaporation above oceans (Byrne and O'Gorman 2018; Brogli et al. 2019). A complex picture of European summer climate change thus emerges (Zappa 2019), in which several robust and well-understood physical mechanisms combine to bring hotter and more arid summers. Still, it remains unclear why the summertime European climate, in comparison to other geographical regions, would be particularly sensitive to soil moisture feedbacks and circulation changes. What makes the region so vulnerable, leading to such unparalleled drying trends in future projections? We provide an answer to this question by putting European summer drying trends in the context of the continent's current climate. First, we show why regions south of 45°N are vulnerable to changes in wet-season precipitation, and we demonstrate that mechanisms responsible for the projected winter precipitation decline in the Mediterranean also come into play during spring, expanding the precipitation decline northwards into Southern Europe and setting the stage for soil moisture depletion into summer. By investigating the detailed vertical structure and meridional distribution of the moisture inflow from the Atlantic, we also show that the summer climate north of 45°N is sensitive to trends in low-level moisture flux along the western coast of the continent. We conclude by offering a comprehensive picture of physical mechanisms of future European summer drying, and discuss their relevance to understand inter-model differences and reduce the associated uncertainty.

Data and methods

Our analysis focuses on a domain (hereafter referred to as the "European domain") that extends from 10°W to 30°E and 40°N to 50°N. The choice of this domain is based on the pattern of future precipitation projections (Figure 7.1c): it covers most of the land area in Europe where models project a robust summer precipitation decline. We focus on three seasons: winter (December-to-February, DJF), spring (March-to-May, MAM) and summer (June-to-August, JJA). The reference for the current climate is ERA5 (Hersbach et al. 2020), the latest of the ECMWF reanalyses. ERA5 is based on an atmospheric model and reanalysis system with a horizontal grid spacing of about 31 km, and is currently available over the 1979-2019 period. Using ERA5, we analyze specific humidity, relative humidity (RH), temperature and winds archived on a regular 1° grid with 17 pressure levels, as well as precipitation (P) and evapotranspiration (E). Horizontal winds and specific humidity are downloaded at a 6-hourly resolution to calculate time-mean horizontal moisture fluxes, while other variables are extracted at monthly resolution only. Because P in ERA5 is a prognostic variable (i.e., not directly assimilated), we also use observation-based P datasets at the monthly timescale: the Global Precipitation Climatology Project version 2.3 dataset (GPCP, 2.5° resolution) (Adler et al. 2003), and the two land-only CRU TS4.04 (0.5° resolution) (Harris et al. 2020) and EOBS (0.25° resolution) (Cornes et al. 2018) datasets. E in ERA5 is also not assimilated, and thus influenced by inevitable biases in P and soil moisture data. Finally, soil moisture observations (top meter) are taken from the ESA CCI Soil Moisture v05.2 combined product (Gruber et al. 2019). Monthly-mean soil moisture values are calculated by averaging all non-missing observations at each grid cell and for each available month. For purposes of comparison, we also analyze top 1 meter soil moisture from ERA5. Climate model output data from CMIP5 models (Taylor et al. 2012) are also analyzed, under the historical (1976-2005) and RCP8.5 (2071-2100) scenarios. We retained all 29 models that provided SLP, P, E and RH at monthly resolution. For analyses involving moisture fluxes, we use a subset of 23 models for which horizontal wind and specific humidity data on pressure levels were available at daily resolution. Similarly, soil moisture data is only available for 19 of the 29 models. A detailed list of the selected models can be found in Table 2.1.

All model output is bilinearly regridded to a common 1° x 1 grid, with land and ocean values regridded separately for surface/1000 hPa variables only. Future changes are defined under RCP8.5 with respect to the 1976-2005 historical average. We normalize them by each GCM's annual- and global-mean temperature change to remove, at first order, the effect of varying climate sensitivity across models. This is important to avoid overestimating inter-model correlations in long-term changes. Given *q(p)* the specific humidity (kg/kg) and $\mathbf{u}(p)$ the horizontal wind (m/s) at pressure p (Pa), at a 6-hourly (in ERA5) and daily (in CMIP5) resolution, we compute the seasonal-mean tropospheric moisture flux as the vertical integral of the seasonal-averaged moisture flux $\overline{q\mathbf{u}}$ between p_0 = 1000 hPa and p_1 = 100 hPa:

$$\mathbf{F} = -\frac{1}{g}\int_{p_0}^{p_1} \overline{q\mathbf{u}}dp \qquad (7.1)$$

where $\overline{\cdot}$ stands for the seasonal average and g = 9.81m/s^2. For the GCM simulations, we also define estimates of future summertime moisture fluxes under a "constant-wind-and-RH" assumption as follows. For each model, pressure level p and grid point, we note $\overline{rh_{hist}}$ the JJA-mean RH in the historical scenario; $\mathbf{u}_{hist}(t)$ and $T_{hist}(t)$ the daily horizontal wind and air temperature in the historical scenario; and $T_{RCP85}(t)$ the daily air temperature in the RCP8.5 scenario. We first sort JJA-only daily T_{RCP85} values in the same order as those of JJA-only T_{hist}. This yields the series $T_{\mathrm{RCP85}}(t')$. Then, we calculate at time step t a future specific humidity value assuming no change in mean seasonal RH as:

$$\hat{q}_{RCP85}(t) = \epsilon \frac{e_s(T_{RCP85}(t'))}{p} \overline{\mathrm{rh}_{\mathrm{hist}}} \qquad (7.2)$$

where $e_s(T)$ is the saturation vapor pressure of water at temperature T and $\epsilon = 0.622$ is the molecular weight ratio of water vapor to dry air. The estimated horizontal moisture flux $\widehat{\boldsymbol{F}}_{RCP85}$ is then obtained by

$$\widehat{\boldsymbol{F}}_{RCP85} = -\frac{1}{g}\int_{p_0}^{p_1} \overline{\hat{q}_{RCP85}\mathbf{u}_{\mathrm{hist}}} dp \qquad (7.3)$$

The seasonal atmospheric water budget for an atmospheric column above area $\mathcal{A}$ can then be written as

$$P - E = -\nabla \cdot \mathbf{F} \qquad (7.4)$$

with P, E and $\mathbf{F}$ the area-integrated seasonal-mean values of precipitation, evapotranspiration and horizontal moisture flux respectively. The two moisture sources for the atmospheric column above area $\mathcal{A}$ are E and the incoming moisture flux F_{in}, defined as the horizontal advection of moisture into the column from all sides:

$$F_{in} = \oint_{\partial \mathcal{A}} -\mathbf{F} \cdot \mathbf{n} \times 1\{\mathbf{F} \cdot \mathbf{n} < 0\} \qquad (7.5)$$

where $\mathbf{n}$ is the unit vector normal to $\partial \mathcal{A}$. We use F_{in} to assess the role of horizontal moisture fluxes against that of E in the seasonal atmospheric moisture budget over the previously defined European domain. Finally, the seasonal surface water budget for area $\mathcal{A}$ is given by

$$P - E = \Delta W_s + R \qquad (7.6)$$

where ΔW_s is the area-integrated change in water storage and R the runoff flowing out of area $\mathcal{A}$.

As in Chapter 2, to which we refer the reader for a more complete description of the methodology, we estimate the impact of upper-tropospheric (300 hPa) wind trends on Northern Hemisphere SLP during MAM with a dynamical adjustment technique based on constructed circulation analogs. For each GCM, analogs of each future MAM 300 hPa zonal wind anomaly field between 2071 and 2100 are identified in that model's historical run, and the corresponding SLP anomalies are linearly combined with optimal weights. The domain used to identify analogues and compute the weights is the latitude band extending from 15° to 65°N. We also make use of the regional simulations discussed in Chapter 2 to assess the impact of reduced rates of warming of near-surface air above the Mediterranean Sea compared to surrounding continents. The simulations extend over a domain covering most of Europe and the Mediterranean Basin, spanning approximately 24-53°N, 20°W-45°E, at a horizontal resolution of 35 km and using 40 vertical levels. They are run with the MIT Regional Climate Model (MRCM). MRCM is based on the Abdus Salam International Centre for Theoretical Physics Regional Climate Model, version 3 (RegCM3) (Pal et al. 2007), and includes several improvements achieved through incorporation of new physical schemes or modification of original schemes (see Chapter 2) and reference therein). Two sets of simulations were conducted (Im et al. 2014): one forced using ERA-Interim as boundary conditions over the 1981-2011 period, and one forced with output from the MPI-ESM-MR model under the RCP8.5 scenario between 2070 and 2100. Each set consists of a reference (unperturbed) experiment (designated by "0C") and a perturbed experiment in which

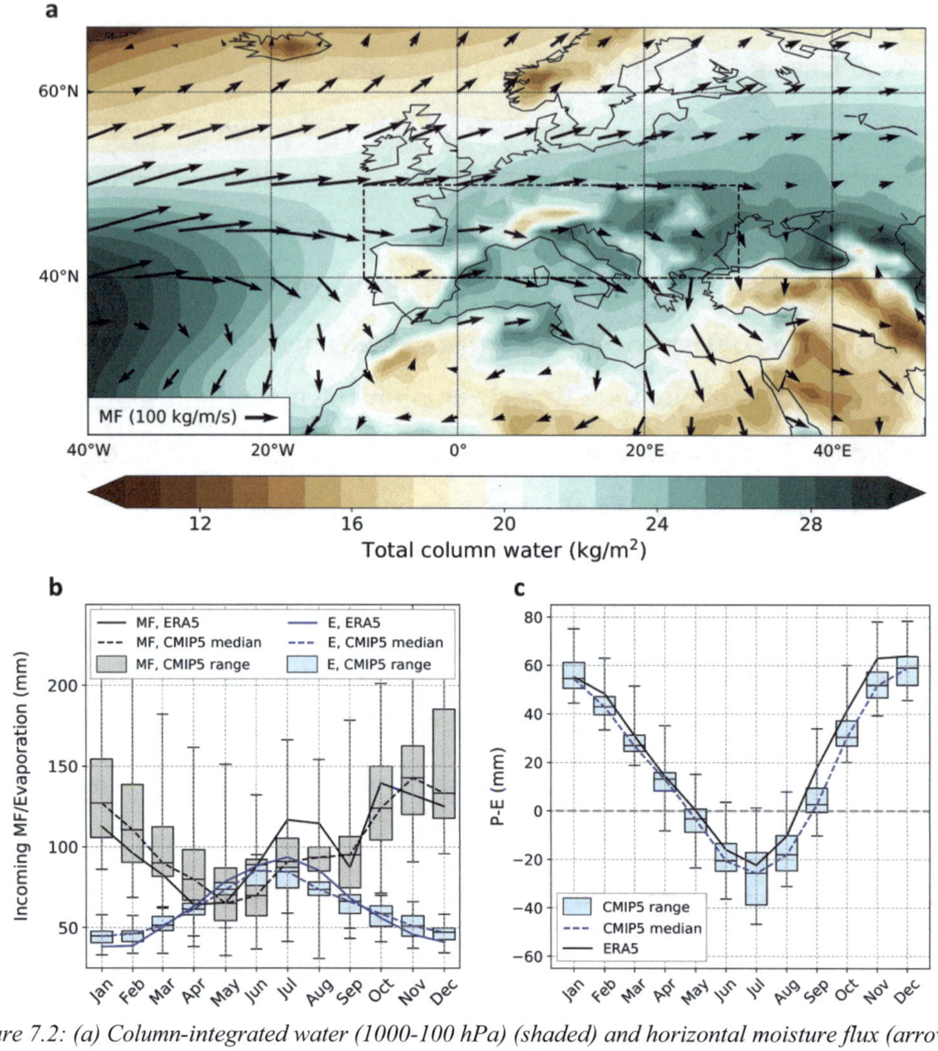

Figure 7.2: (a) Column-integrated water (1000-100 hPa) (shaded) and horizontal moisture flux (arrows) over Europe during JJA, in the ERA5 reanalysis (1979-2019). (b) Annual cycle in incoming moisture flux Fin (black/gray shading) and surface (land + ocean) evaporation (blue) within the European domain (10°W-30°E/40°-50°N), in ERA5 (solid lines) and across CMIP5 models (boxes and dashed lines). (c) Annual cycle in P-E (land-only) over the European domain, in ERA5 (solid black) and across CMIP5 models (boxes, dashed blue line). In (b,c), whiskers extend over the full inter-model spread.

Mediterranean SSTs are uniformly warmed by 1.5°C year-round (designated by "+1.5C"). The impact of the relative SST cooling on the regional climate can then be estimated by subtracting the results from the +1.5C simulation to those of the 0C simulation. The MPI experiments are intended to assess the impact of a relative Mediterranean cooling in a climate change context. Since any GCM presents a number of biases in its background climate, which may modulate the impact of the relative cooling, the ERA experiments thus allow us to assess the circulation impact of a relative cooling in the current Mediterranean climate with no land warming.

Europe's summer moisture budget in the current climate

To understand the large-scale historical context of summer drying projections over Europe, we begin by showing on Figure 7.2a tropospheric horizontal moisture fluxes and total column water (TCW) during the summer (June-to-August, JJA) season. The European domain stands out as a local TCW maximum, particularly its eastern part (Balkans). TCW in regions to the north of the domain, polewards of 55°N, is about as low as over the Sahara Desert to the south (15-20 kg/m^2). To the west, over the Atlantic Ocean, TCW is also comparatively lower over an elongated coastal band between 20°W-0°. The summer horizontal moisture flux is characterized by a narrow, westerly low-level jet extending between 45°N and 55°N, and stretching from the North Atlantic into Northern Europe, up to 20-30°E. As a consequence, the incoming moisture flux into the European domain F_{in} is largely westerly, with the western border of the domain accounting for 80% of the magnitude of F_{in}.

How does Europe maintain its humidity under such conditions? On the one hand, atmospheric moisture fluxes advect moisture onto the continent, but

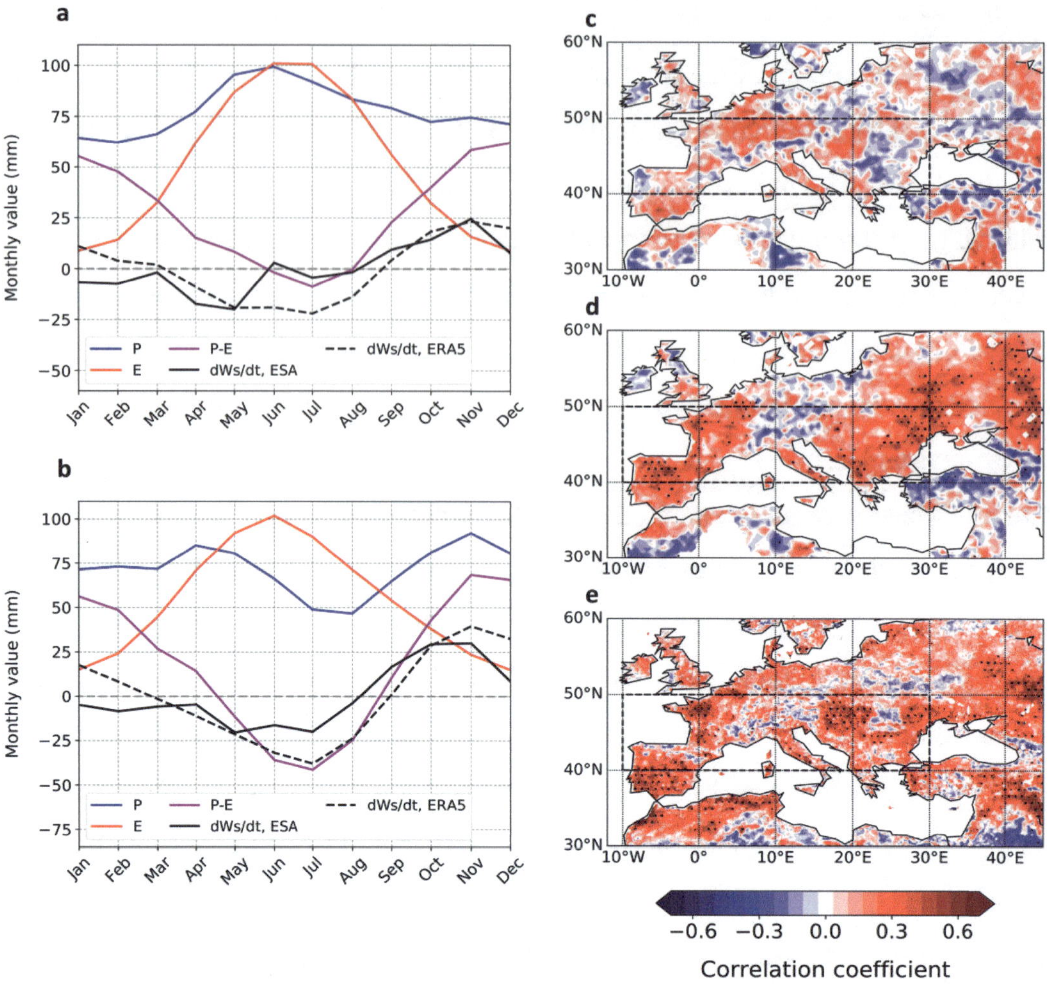

Figure 7.3: (a) Annual cycle in land-only P (blue), E (red), P-E (purple) and monthly soil moisture changes (black) for the southern half of the European domain (10°W-30°E/40°-45°N). All data is from ERA5 except soil moisture (solid black) from the ESA dataset. (b) Same as (a), but for the northern half of the domain (10°W-30°E/45°- 50°N). (c-d) Pointwise correlation coefficient between (c) DJF and (d) MAM P (from EOBS dataset) and June soil moisture (from ESA dataset). (e) Pointwise correlation coefficient between May soil moisture (from ESA dataset) and JJA E (from ERA5). Stippling in (c-e) indicates significance at the 10% level, after correcting for the False Discovery Rate according to Wilks (2016).

mainly northwards of 45°N (Seager et al. 2014). In addition, the moisture supply coming directly from surface (land and ocean) evapotranspiration (E) within the domain makes up a substantial fraction of incoming moisture into the lower troposphere (Figure 7.2b). E reaches its minimum in January and its maximum in July, and its share in the total moisture supply (F_{in}+E) goes from ≈30% during winter to ≈50% during summer, a phenomenon generally well reflected in CMIP5 models (Figure 7.2b). In ERA5, F_{in} is about as large in summer as in winter, despite higher specific humidity, since the storm track is displaced polewards. In CMIP5 models, however, the summer F_{in} tends to be smaller than in winter, possibly a consequence of the zonal bias in the North Atlantic storm track in GCMs. E is maximal during summer, even exceeding P, which leads to negative P-E (Figure 7.2c) from June to August. Negative P-E can only be sustained by the depletion of soil and surface water storage (ΔW_s, equation 7.6). The latter is replenished from October to April, when P exceeds E.

The analysis of Figure 7.2 concerns the European domain as a whole. However, the two halves of the domain (northern, from 45-50°N, and southern, from 40-45°N) behave differently (Figure 7.3a,b). During summer, the northern half stands under the influence of the mid-latitude storm track, which advects significant amounts of moisture and results in substantial P (its annual maximum is reached in June) (Figure 7.3a). Consequently, while E is high, P-E remains close to zero from June to August, and soil moisture in the ESA data hardly decreases. Still, ERA5 is in disagreement as it points towards a substantial soil moisture decline in the same period (down to -20 mm in June). Soil moisture increases again in September as P-E becomes positive. By contrast, southwards of 45°N, P drops during summer, just as E reaches its maximum (Figure 7.3b). P-E is

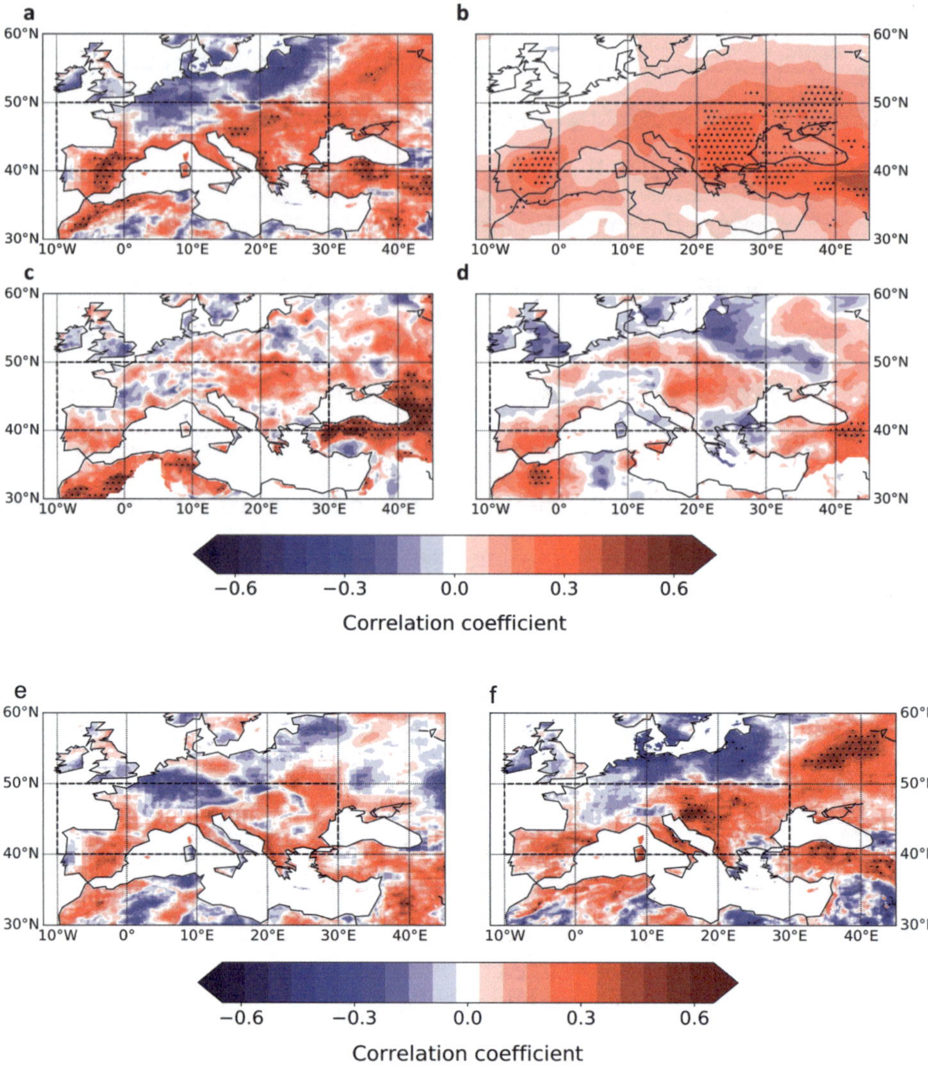

Figure 7.4: (a) Pointwise correlation coefficient between December-to-May P-E and JJA 850 hPa relative humidity (ERA5 data, 1979-2019). (b) CMIP5 multi-model mean pointwise correlation between December-to-May PE and JJA 850 hPa relative humidity, in the historical (1976-2005) simulations. (c,d) Pointwise correlation coefficient between December-to-May P in (a) EOBS and (f) CRU, and JJA 850 hPa relative humidity in ERA5 (1979-2019). Stippling in (b) indicates agreement between at least 80% of CMIP5 models on the sign of the correlation, and in all other panels indicates significance of the correlation at the 10% level, after correcting for the False Discovery Rate according to Wilks (2016).

thus strongly negative from May to September, which leads to soil moisture depletion. ESA and ERA5 data are in agreement on the sign of soil moisture changes, but the magnitude is about twice as large in ERA5 as in ESA. Both datasets also agree that soil moisture changes during summer are more negative than in the northern half. Again, soil moisture increases from September onwards, when P-E becomes positive.

This suggests that, south of 45°N, a larger fraction of summer E (and runoff) es from soil moisture depletion compared to regions north of 45°N. The relationship between winter/spring P, P-E, soil moisture anomalies and summer evaporation also emerges in the analysis of inter-annual variability (Figure 7.3c,d,e). While P in DJF is not significantly correlated to June soil moisture (Figure 7.3c), it is for MAM in many parts of the European domain: the Iberian Peninsula, large parts of France, Italy and the Balkans (Figure 7.3d). May soil moisture anomalies in turn significantly impact JJA E over the same regions (Figure 7.3e). Consequently, P-E during spring (and to a lesser extent winter), which replenishes soil moisture levels before the summer, is positively correlated to summer 850 hPa RH over much of the southern half of the European domain (Figures 4a). Polewards of 45°N, the connection between winter and summer conditions appears reversed, though not significant. A similar pattern is also present in observation-based P datasets (Figure 7.4c,d) but with weaker magnitude and no statistical significance, which might be due to the fact that P is less relevant than P-E for the soil moisture budget. The correlations between spring P-E summer and 850 hPa RH are significant over much of the Iberian Peninsula and locally over the Balkans, regions where summer E is most reliant on soil moisture accumulated during the cold season. While correlation

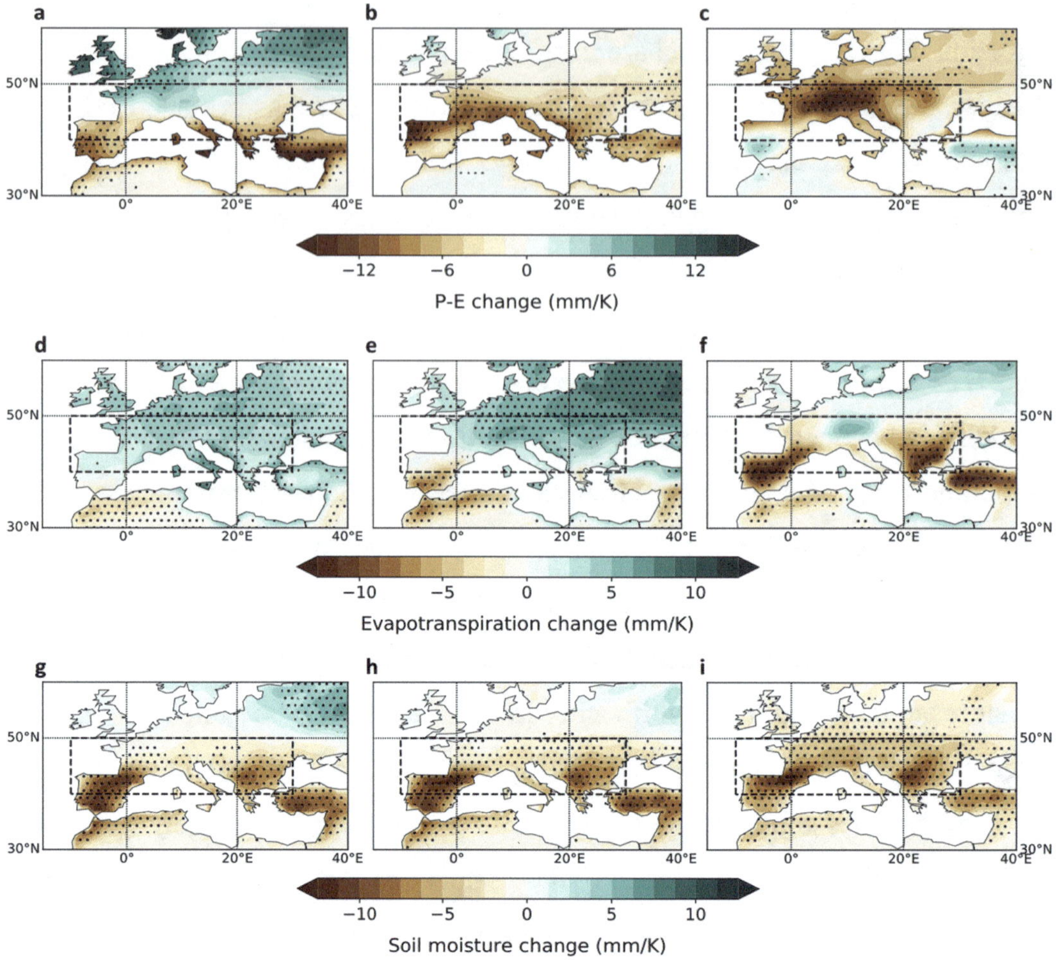

Figure 7.5: (a-c) CMIP5 multi-model mean change (2071-2100 minus 1976-2005) in (a-c) P-E, (d-f) E and (g-i) soil moisture (top 1 meter) over Europe (land only), in (a,d,g) DJF, (b,e,h) MAM and (c,f,i) JJA. Stippling indicates agreement by at least 80% of models on the sign of the change.

significance is limited in observations, CMIP5 models reproduce the overall correlation pattern over Southern Europe, with strong inter-model agreement over the Iberian Peninsula and the Balkans (Figure 7.4b).

This increases our confidence in the physical interpretation of the observation-based correlations. During summer, the European domain is therefore dependent on wet-season water storage and on the narrow low-level Atlantic moisture jet to maintain its humidity. As such, any change in either of these moisture sources driven by climate change can impact summer moisture availability over Europe and, consequently, precipitation. Based on this perspective, we consider in the following each moisture source in turn, look at its projected trends under future climate scenarios, and explain how these trends contribute to projections of summer drying over Europe.

Wet season precipitation decline and impact on European summer climate

Changes in the surface water balance: winter, spring and summer

Climate models have long projected a significant decline in precipitation during winter and spring over Southern Europe and the Mediterranean Basin (Seager et al. 2014; Lionello and Scarascia 2018). In winter, absolute changes are most negative over Northwestern Africa, the southern Iberian Peninsula and along the Eastern Mediterranean coastline (see Chapter 2), whereas over Northern Europe precipitation is expected to increase substantially (Figure 7.1a). During spring, the decline in precipitation shifts northwards into Southern Europe, extending up to 45°N, while from 45 to 50°N precipitation is projected to remain roughly constant (Figure 7.1b). These patterns are reflected in P-E projections. During winter, the P-E decline is essentially limited to the Mediterranean coastline, and is robust in the southern half of our domain (Figure 7.5a). During spring, it extends northwards into Western and Central Europe, up to 50°N (Figure 7.5b). In summer, the decline progresses further northwards: it is mainly concentrated at latitudes higher than 45°N, while P-E is projected to remain constant, or even to increase slightly, in the regions to the south (Figure 7.5c).

In winter and spring, the projected decreases in P-E over the western half of the European domain (Iberian Peninsula and Italy in DJF/MAM, France in MAM) are mainly driven by negative P trends, since projections show relatively constant or increasing E in these regions (Figure 7.5d,e). In the eastern half, however (Balkans), the P decline is mostly restricted to Greece; higher north it is the increase in E that explains the negative P-E trend. In summer, the P decline between 40-45°N is roughly compensated by a decline in E, particularly strong over the Balkans and the Iberian Peninsula (5-f), leading to the weak P-E trends of Figure 7.5c. By contrast, E remains relatively unchanged north of 45°N, and it is the decrease in P that explains the negative P-E change. In terms soil moisture, changes within our domain are mainly negative (Figure 7.5g,h,i). Large year-round declines are projected for the Iberian Peninsula and the Balkans. For the rest of the domain, the decline mainly occurs during summer, in parallel with that in P-E (5-c). Soil moisture trends during winter and spring, though less robust, also point to a slight decline as well, despite P-E remaining about constant there (5-a), a likely consequence of the summer decline whose impact is felt throughout the year.

Mechanisms of spring precipitation and P-E change

The robust precipitation decline during winter, centered over the Southwestern and Eastern Mediterranean, is the result of an increase in average SLP over the Mediterranean Basin (Seager et al. 2014; Chapter 2). Higher Mediterranean SLP

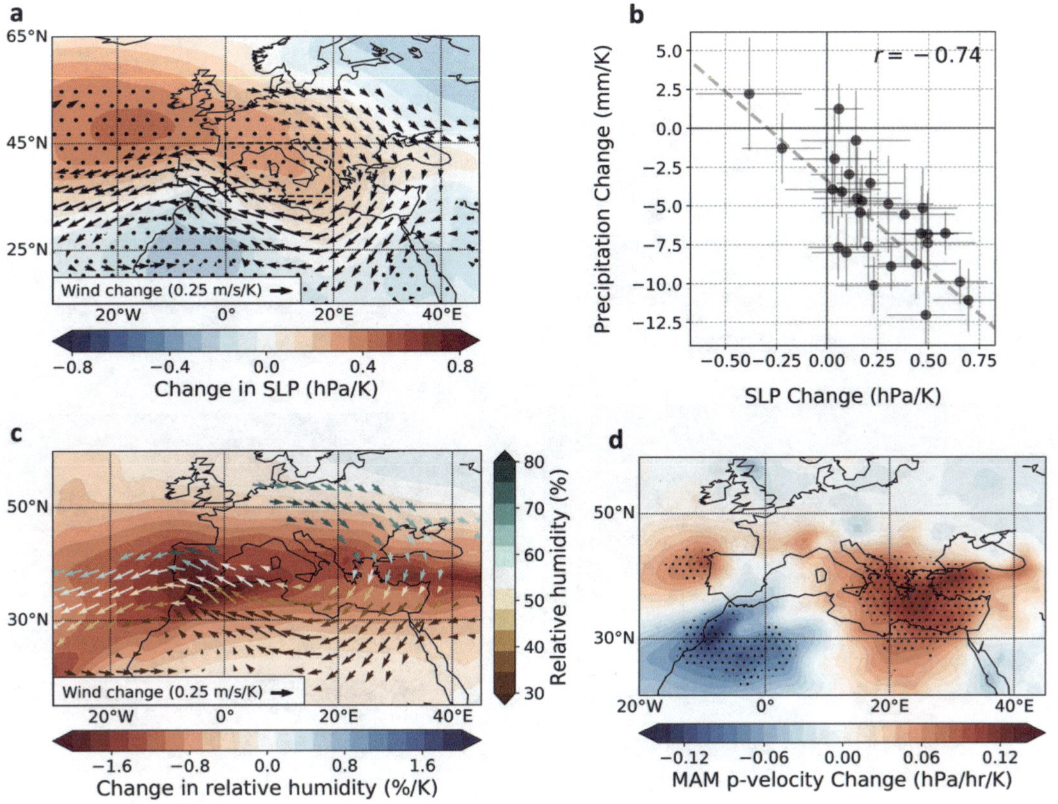

Figure 7.6: (a) CMIP5 multi-model mean change (2071-2100 minus 1976-2005) in MAM SLP (shaded) and 850 hPa winds (arrows). (b) Change in MAM Mediterranean (10°W-40°E/35°-45°N) precipitation and SLP (0°-20°E/35°-45°N, see box on (a)) across CMIP5 models, with 95% confidence intervals and least-squares best fit. (c) CMIP5 multimodel mean change in MAM 850 hPa relative humidity (shaded) and 850 hPa winds (arrows), where wind arrows are colored according to mean historical relative humidity at each point. (d) CMIP5 multimodel mean change in MAM 500 hPa pressure velocity. In (a,d) stippling indicates agreement by at least 80% of models on the sign of the change. In all panels changes are taken under the RCP8.5 scenario and are normalized by each model's global-mean projected temperature change.

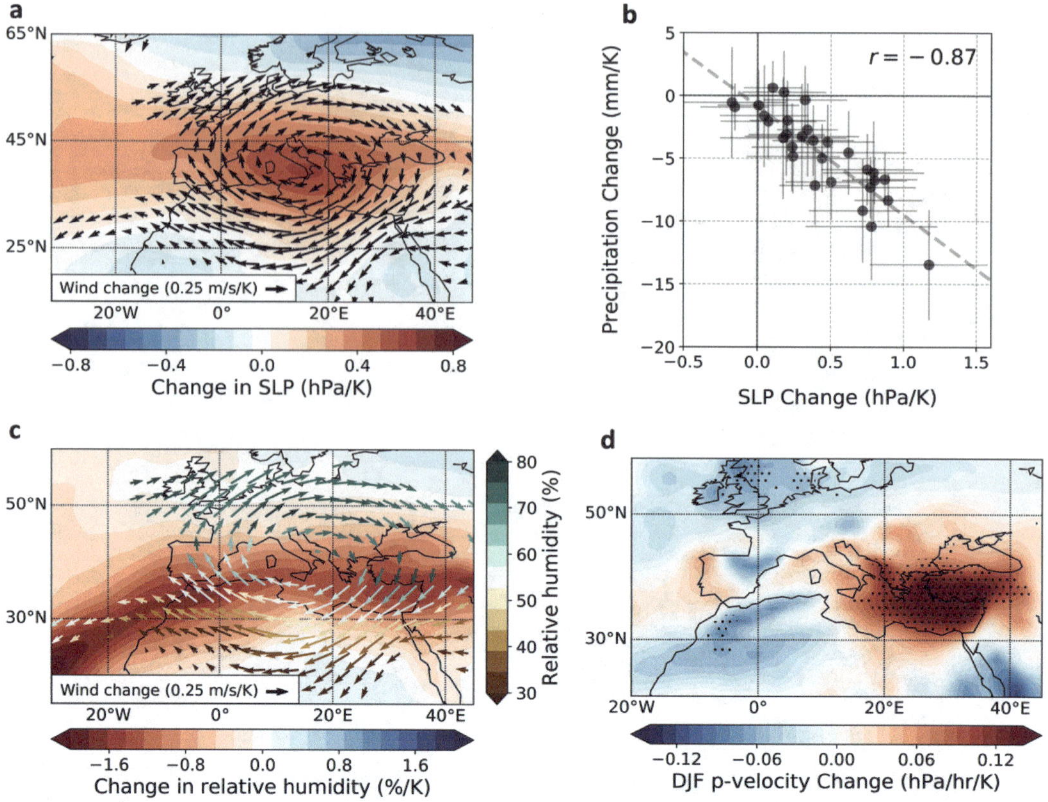

Figure 7.6.2: (a) CMIP5 multi-model mean change (2071-2100 minus 1976-2005) in DJF SLP (shaded) and 850 hPa winds (arrows). (b) Change in DJF Mediterranean (10°W-40°E/30°-45°N) precipitation and SLP (0°-30°E/32°-48°N) across CMIP5 models, with 95% confidence intervals and least-squares best fit. (c) CMIP5 multimodel mean change in DJF 850 hPa relative humidity (shaded) and 850 hPa winds (arrows), where wind arrows are colored according to mean historical relative humidity at each point. (d) CMIP5 multimodel mean change in DJF 500 hPa pressure velocity. In (a,d) stippling indicates agreement by at least 80% of models on the sign of the change. In all panels changes are taken under the RCP8.5 scenario and are normalized by each model's global-mean projected temperature change.

indeed drives anomalous northerlies and the associated cold (and dry) advection over the Eastern Mediterranean which, following the energy balance, triggers a diabatic warming and a reduction in precipitation through anomalous subsidence.

To the west, the SLP anomaly generates anomalous dry advection from the Sahara Desert towards Morocco and the southern Iberian Peninsula, in addition to weakening the Atlantic moisture supply, which reduces RH in the lower troposphere and decreases convective instability (see Chapter 4). The pronounced SLP increase is itself a response to shifts in the upper-tropospheric planetary circulation and to the regional-scale warming contrast between the Mediterranean and surrounding land (Chapter 2). The strengthening of upper-tropospheric zonal winds, linked to enhanced warming at upper levels in the tropical troposphere, lengthens the wavelength of stationary planetary waves (Simpson et al. 2016; Wills et al. 2019). This leads to a robust anticyclonic anomaly over the Mediterranean, which, due to zonal asymmetries in the vertical structure of stationary waves, is transmitted down to the surface with limited dampening (see Chapter 3). Additionally, the surface heating anomaly due to reduced rates of warming of Mediterranean SSTs compared to land also leads to higher SLP over the basin. During spring, Southern Europe and the Mediterranean are also characterized by significant positive SLP trends and associated anticyclonic circulation (Figure 7.6a).

While winter trends are characterized by a strong anomalous ridge over the Mediterranean, European SLP trends during spring are weaker (roughly 1 hPa compared to 2 hPa) and displaced to the west, where they connect with an area of robust enhanced SLP over the North Atlantic. Still, as is the case during winter (Chapter 2), Southern European and Mediterranean precipitation trends across

CMIP5 models are primarily connected to SLP changes over Southern Europe (southwards of 45°N), and not over Northwestern Europe and the Atlantic Ocean (Figure 7.6b), even though the latter are also robust. The relationship between spring Mediterranean (10°W-40°E/35-45°N) precipitation projections and the SLP increase over Southern France and Italy in particular is very strong and explains half of inter-model scatter in precipitation projections ($r^2 = 0.55$). The reason, as in winter, can be found in the circulation anomaly triggered by such SLP trends. The robust SLP increase centered around 10°E/40°N generates a substantial anticyclonic anomaly associated with anomalous southeasterlies and dry advection over Southwestern Europe (Figure 7.6c), and anomalous northerlies, cold and dry advection, as well as subsidence, over Southeastern Europe and Turkey (Figure 7.6d).

Comparing spring (Figure 7.6) to winter (Figure 7.6.2), the P and P-E declines shift northwards during spring, particularly over the Western Mediterranean. This shift is partly related to a similar northward drift of the anomalous SLP ridge from the central to the northwestern Mediterranean (Figure 7.6a). Anomalous SLP gradients over the Western Mediterranean are weaker during spring, but low-level wind trends are of a similar magnitude to those projected for winter (1-1.5 m/s). In addition, the projected low-level RH decline is stronger than during winter (Figures 7.6c and 7.6.2c). This may result from the fact that the humidity difference (both absolute and relative) between the Sahara Desert and the Iberian Peninsula is larger during spring, and also from the storm track and associated moisture transport being more polewards during spring. To the east, anomalous subsidence trends are noticeably weaker in MAM than in DJF, possibly a consequence of precipitation being smaller in this region during

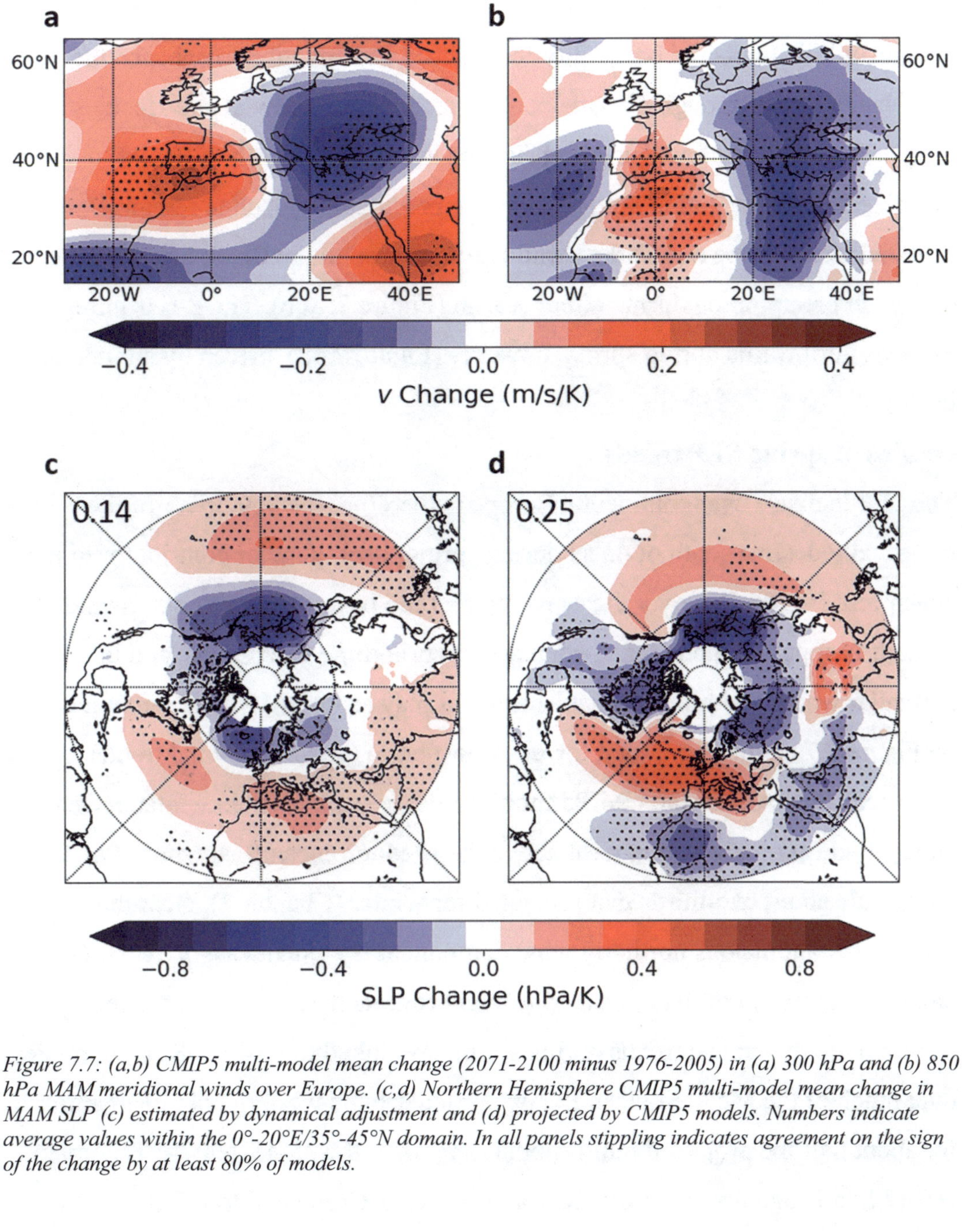

Figure 7.7: (a,b) CMIP5 multi-model mean change (2071-2100 minus 1976-2005) in (a) 300 hPa and (b) 850 hPa MAM meridional winds over Europe. (c,d) Northern Hemisphere CMIP5 multi-model mean change in MAM SLP (c) estimated by dynamical adjustment and (d) projected by CMIP5 models. Numbers indicate average values within the 0°-20°E/35°-45°N domain. In all panels stippling indicates agreement on the sign of the change by at least 80% of models.

spring. Indeed, lower precipitation means less latent heating of the atmospheric column, which limits the possibility for reduced precipitation to enhance the subsidence signal through diabatic cooling feedbacks (Rodwell and Hoskins 2001). The location of enhanced subsidence is not significantly different in spring compared to winter, and in both seasons the precipitation decline remains centered over Greece and Turkey while regions around the Black Sea see little change in precipitation during either season (Figure 7.1a,b). The P-E decline expands northwards during spring, however (Figure 7.5b), driven by an increase in E.

Origins of spring SLP trends

The SLP increase over continental Europe is therefore the key driver of the projected P decline south of 45°N during spring, itself a major contributor to the concurrent P-E decline over the same region (and beyond, over Southern and Eastern Europe). To understand what drives springtime SLP trends, and in particular whether similar mechanisms are at play as during winter, we first show on Figure 7.7a,b the upper- and lower-tropospheric meridional wind trends projected for MAM by the CMIP5 multi-model ensemble. As in winter, a clear anticyclonic anomaly occurs right above the Mediterranean Basin, but of a magnitude about two-thirds that projected for winter (Chapter 3). Near the surface, the anomalous northerly wind component is evident east of 20°E. The empirical approach of dynamical adjustment reveals that, similar to winter, upper tropospheric dynamical trends during spring are linked to positive SLP anomalies over much of the Mediterranean Basin, particularly its western half, accounting for about half the projected multi-model trend over the 35-45°N/0-20°E domain (+0.12 hPa/K against +0.23 hPa/K, Figure 7.7c,d). Compared to DJF,

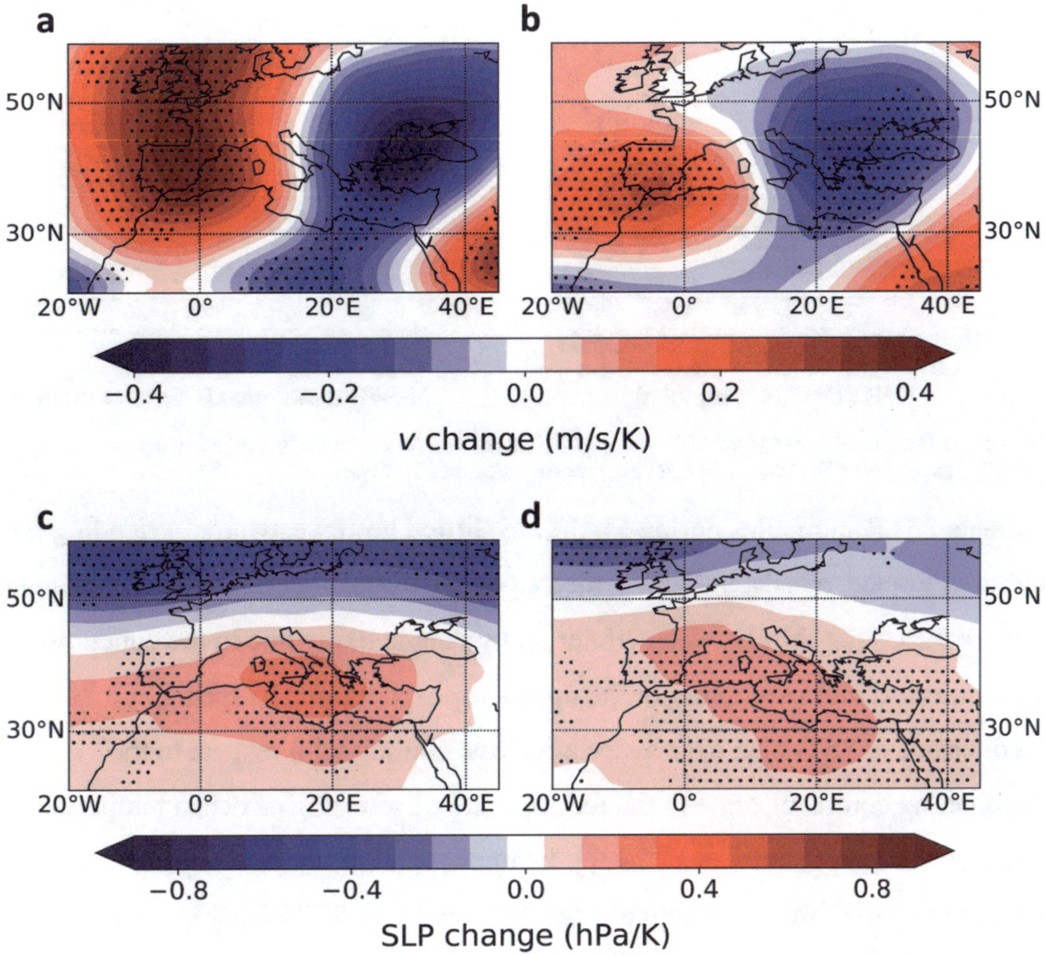

Figure 7.8: (a,b) CMIP5 multi-model mean change (2071-2100 minus 1976-2005) in (a) DJF and (b) MAM 300 hPa meridional winds over Europe. (c,d) CMIP5 multi-model mean change in (c) DJF and (d) MAM SLP estimated by dynamical adjustment. In all panels changes are normalized by each model's global-mean projected temperature change, and stippling indicates agreement on the sign of the change by at least 80% of models.

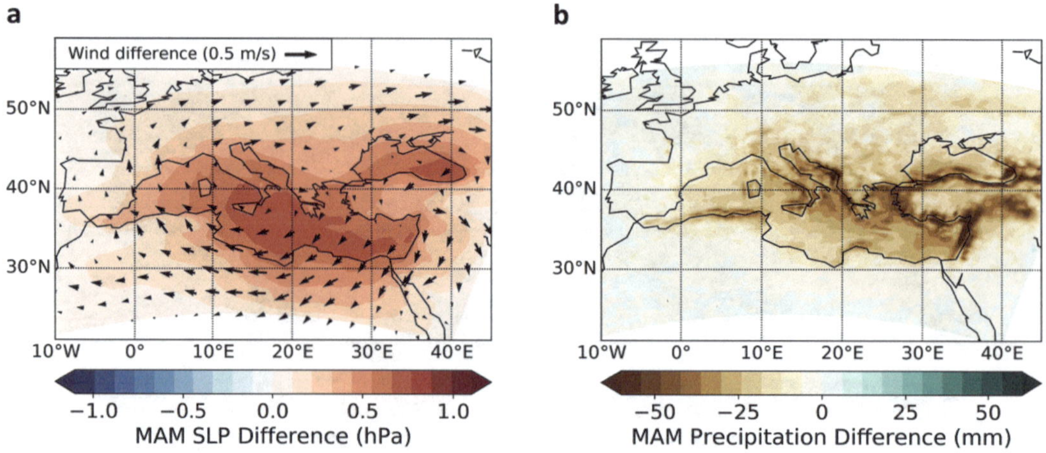

Figure 7.9: Difference in MAM (a) SLP and 850 hPa winds, and (b) precipitation, between the "0C" and "+1.5C" simulations ("0C" minus "+1.5C", average of ERA and MPI runs).

estimated SLP anomalies during MAM are shifted northwestwards, extending into southern France, consistent with projected upper-tropospheric circulation changes (Figure 7.8). This shift between winter and spring is likely due to the concurrent poleward drift and weakening of the upper-tropospheric jet. In spring, Mediterranean SSTs also experience a relative cooling with respect to the surrounding continents, due to the reduced rate of warming of ocean temperatures compared to land (Sutton et al. 2007). Its impact on the regional circulation, as estimated by the difference between the "0C" and "+1.5C" MRCM experiments, is a basin-wide SLP increase, reaching a maximum of 0.75 hPa over the central Mediterranean (20°E), and averaging +0.5 hPa over the previously defined South European domain (35-45°N/0-20°E) (Figure 7.9a). The wind response is essentially confined to the southeastern and southern Mediterranean, with very weak changes over Southwestern Europe (≤0.1 m/s). It is also associated with a broad decrease in precipitation (Figure 7.9b), mostly eastwards of 10°E, extending into the Balkans, where it amounts to about 10% of seasonal precipitation. In keeping with the previous discussion, very little difference in precipitation is seen over Southern France and the Iberian Peninsula, where winds are roughly the same in the perturbed and unperturbed experiments.

Overall, the two mechanisms seem to explain most of the SLP change projected by CMIP5 models over Southern Europe. Compared to winter, the spring SLP anomalies linked to each mechanism appear to be displaced westwards (Chapter 2), consistent with CMIP5 projections. It is important to note that neither mechanism can account for the robust SLP increase projected west of France and the British Isles (Figure 7.6a). This positive SLP anomaly is generally thought to be the response to reduced SST warming over the North Atlantic (Haarsma et al. 2015; Gervais et al. 2019). While this SLP feature is not expected to have much influence over Southern European and Mediterranean precipitation trends during spring, it may still matter by preventing the strengthening of mid-latitude westerlies over Northwestern Europe that would be expected from the SLP anomalies from Figures 7.7c and 9a. Indeed, Figure 7.6a shows the lack of a robust wind response over Northwestern Europe, an important difference with winter projections (Chapter 2). In winter, a strengthening in westerlies and in the associated moisture transport likely contributes to the large increase of precipitation projected for Northern Europe, down to 50°N. Its absence during spring may therefore be linked to the much weaker and less robust precipitation signal over Northern Europe, contributing to the overall stagnation in P-E trends (Figure 7.5b).

Moisture flux changes and impact on European summer moisture

Analysis of projected moisture flux trends

Atmospheric circulation changes during winter and spring lead to decreases in P

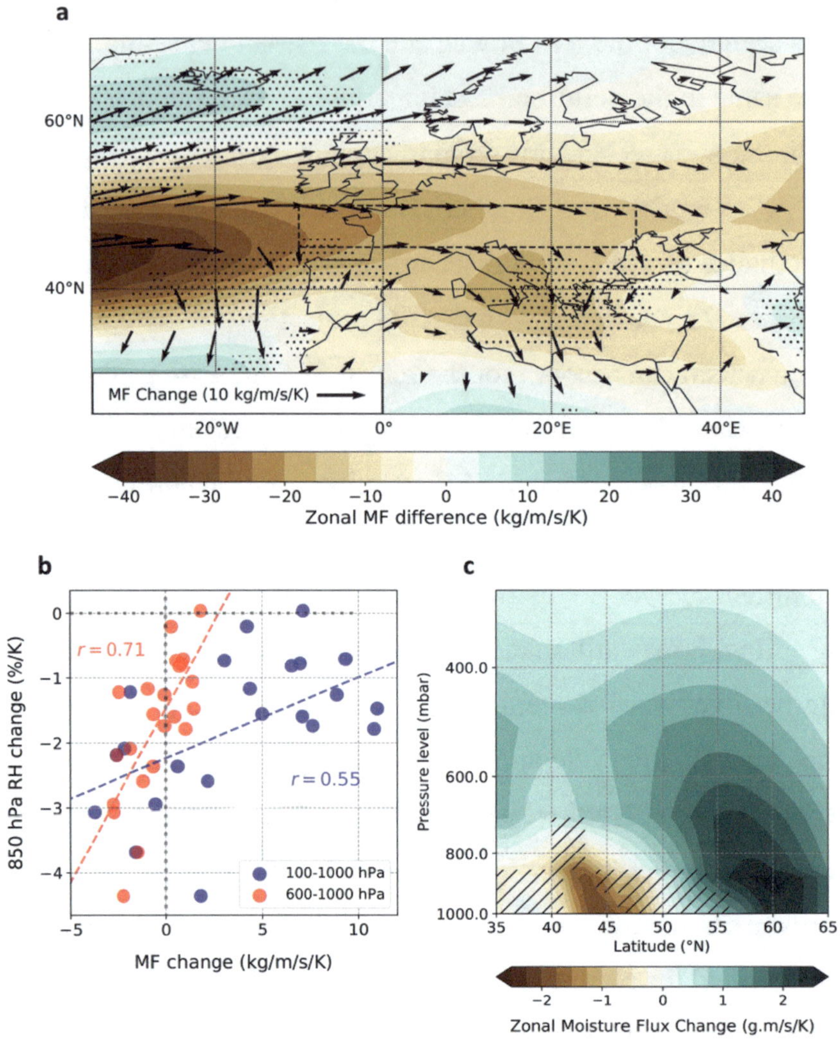

Figure 7.10: (a) CMIP5 multi-model mean change (2071-2100 minus 1976-2005) in JJA vertically-integrated (1000-100 hPa) horizontal moisture flux (arrows), and mean difference between the projected change in zonal moisture flux and the corresponding "constant-wind-and-RH" estimates (see methods) (shaded). (b) JJA change across CMIP5 models in Western European vertically-integrated (1000-100 hPa, blue, and 1000-600 hPa, red) zonal moisture flux at 10 W, averaged bewteen 45°and 50°N, against projected change in 850 hPa RH over the 10°W-30°E/45°-50°N domain (see box in (a)). (c) CMIP5 multi-model mean change in JJA zonal moisture flux at 10°W on pressure levels. Dashed lines in (c) indicate lack of agreement on the sign of the change by at least 80% of models.

and P-E south of 45°N within the domain, setting the stage for a summertime P decline through soil moisture-rainfall feedbacks. North of 45°N, however, winter and spring trends in P-E are weak (Figure 7.5a,b), while summer P and low-level RH still exhibits sharp negative trends (Figure 7.1c,d). In addition, summer E remains approximatively constant in future projections (Figure 7.5f), which is inconsistent with a strong role for land-atmosphere coupling in the P decline. To understand summer P and RH trends, we analyze future horizontal moisture fluxes, particularly along the western border of the European domain (10°W/40-50°N), where the majority of moisture advection into the domain occurs (Figure 7.2a). The vertically-integrated moisture flux increases on average in future projections, leading to enhanced zonal moisture advection into Europe (Figure 7.10a). However, we find, first, that the moisture flux increase along the western border of the domain is lower than what would be expected under an assumption of fixed circulation and RH (Figure 7.10a); and second, that at low levels (below 800), the zonal moisture flux along that same border even decreases in absolute terms (Figure 7.10c) – though the trend is not robust southward of 45°N. Northwards of 60°N, by contrast, zonal moisture fluxes increase across the depth of the troposphere, slightly beyond the fixed-circulation/fixed-RH estimates (Figures 7.10).

The weak increase in vertically-integrated zonal moisture advection has direct implications for the summertime moisture budget over the European domain. Over the northern half of the European domain (see box in Figure 7.10a), where we expect moisture flux trends to matter most for changes in the moisture budget, 850 hPa RH trends are found to be significantly connected to changes in vertically-integrated zonal moisture inflow at 10°W ($r^2 = 0.3$, Figure 7.10b). The

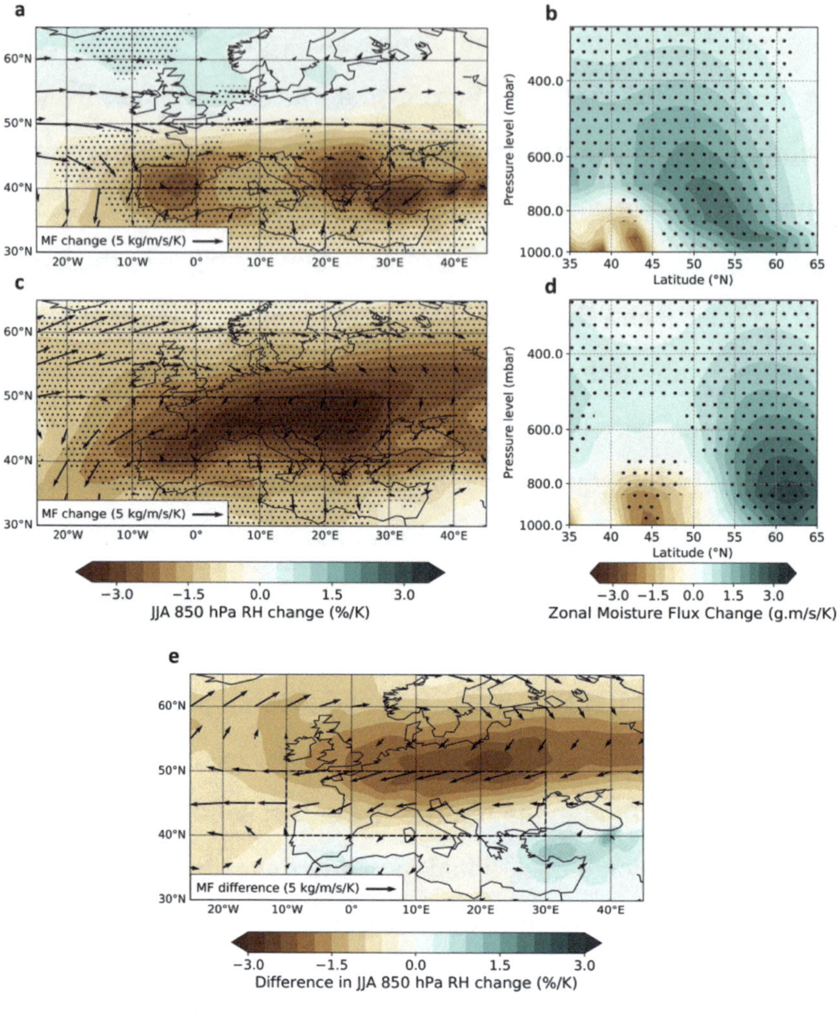

Figure 7.11: (a,b) Change in JJA (a) mean 850 hPa RH (shaded) and 925-600 hPa moisture flux (arrows), and (b) zonal moisture flux averaged bewteen 10°W-0°, for the six CMIP5 models with the most positive moisture flux change in Figure 7.10b. (c,d) Same as (a,b) but for the six models with the most negative moisture flux change. (e) Difference between (c) and (a). In (a-d), stippling indicates agreement on the sign of the change by all the models of the respective group.

correlation is even stronger when restricting the moisture flux to the 1000-600 hPa layer ($r^2 = 0.5$). Inter-model spread in projections remains large, however; 6 GCMs (a fourth of analyzed models) even project an absolute decline in vertically-integrated zonal moisture inflow at 10°W. Figure 7.11 shows 850 hPa RH and moisture flux projections for the 6 models with the most positive (panels a,b) and the most negative (panels c,d) moisture flux change in Figure 7.10b. Clearly, the outlook for regions north of 50°N is very different between these two groups of models (Figure 7.11e). Southern Europe (including the southern half of our domain, below 45°N), on the other hand, experiences sharp drying in either case, particularly over the Iberian Peninsula, the Balkans and Turkey. These are precisely the regions where our previous results suggest that land-atmosphere coupling dominates both for inter-annual and long-term climate variability. The difference in RH trends between the two model groups confirms that uncertainties in moisture flux changes play only a limited role in RH projections (Figure 7.11e).

Role of dynamical trends

Future moisture flux trends are determined by changes in specific humidity and circulation. The fact that zonal moisture fluxes between 40 and 50°N increase less than expected, assuming fixed circulation and RH, implies a decline in RH and/or a change in atmospheric circulation. Though robustly negative, RH trends west of 10°W are weak (Figure 7.1d). Thus, while they undoubtedly contribute to the decline in zonal moisture fluxes relative to the "fixed-circulation, fixed-RH" baseline, they are unlikely to fully account for it. It is also unclear to what extent the RH decline west of 10°W is related to the concurrent decline over the European continent. Low-level circulation trends, on the other hand, clearly show anomalous easterlies over the 30°W-0°/40-45°N domain – where zonal moisture

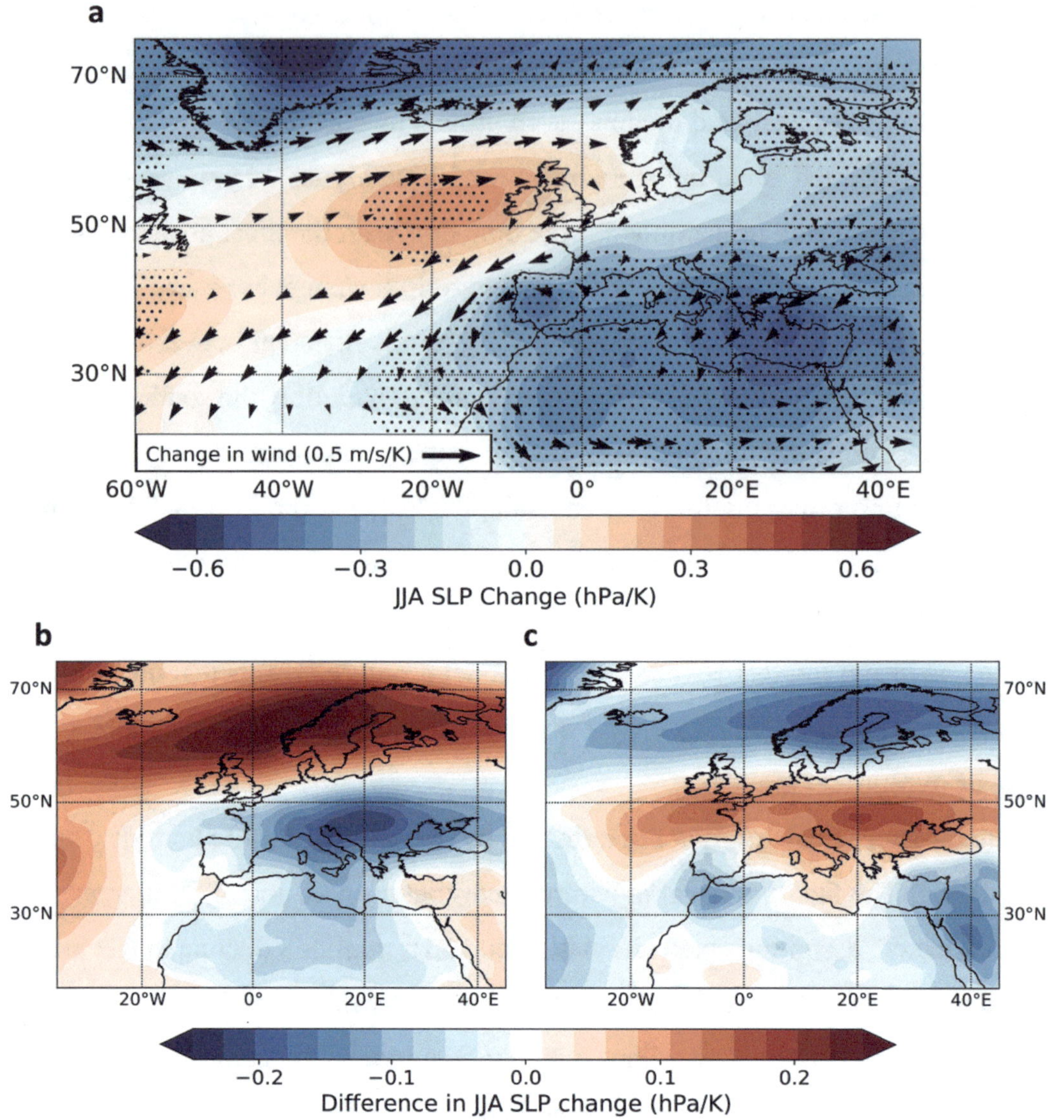

Figure 7.12: (a) CMIP5 multi-model mean change (2071-2100 minus 1976-2005) in JJA SLP (shaded) and 850 hPa winds (arrows), normalized by each model's global-mean projected temperature change. Stippling indicates agreement on the sign of the change by at least 80% of models. (b-c) Difference in JJA SLP change relative to the CMIP5 multi-model mean for the six models with the most (b) negative and (c) positive moisture flux change (Figure 7.10b).

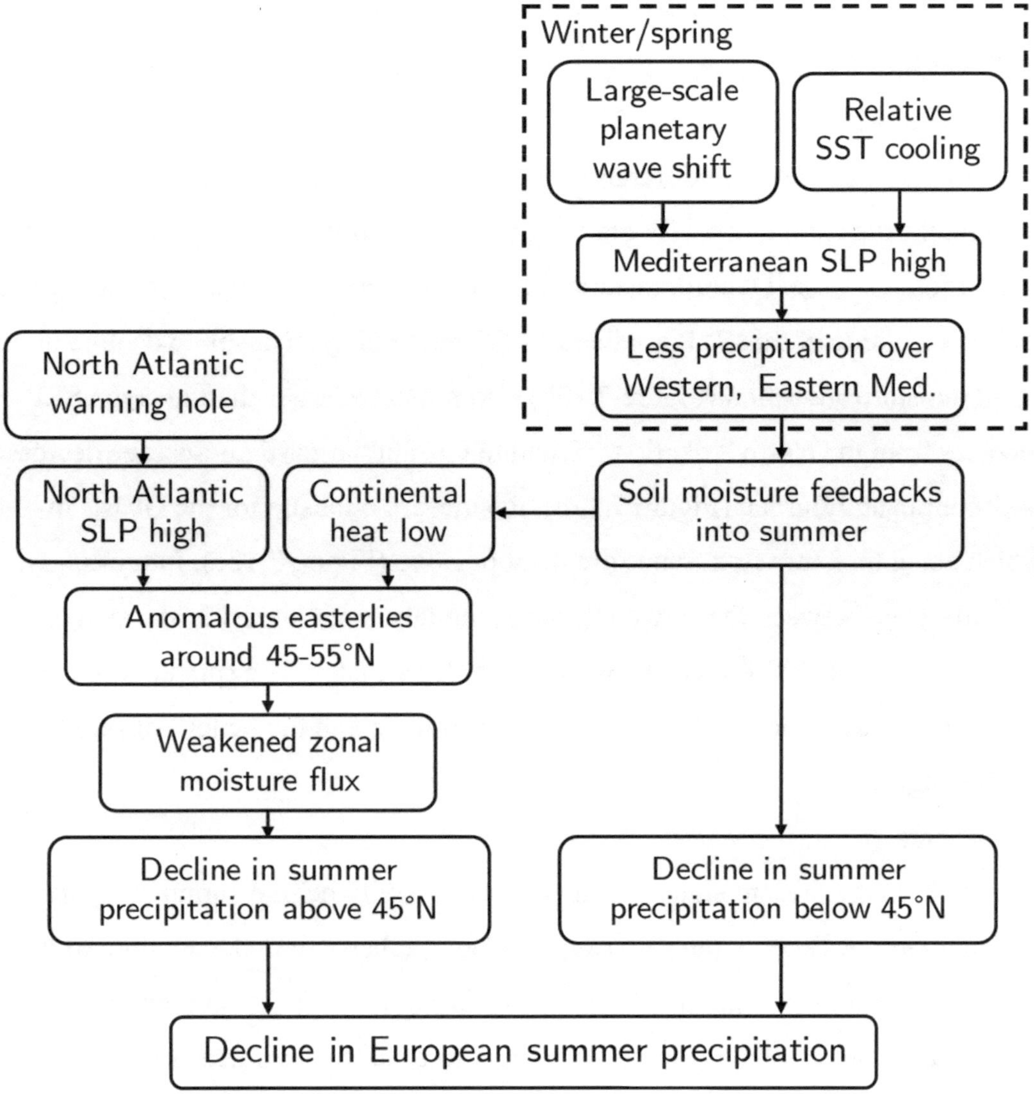

Figure 7.13: Summary of physical mechanisms explaining the projected summer European drying.

fluxes are projected to decline (Figure 7.10c) – and anomalous westerlies northwards of 55°N (Figure 7.12a). Both are linked to positive SLP anomalies west of the British Isles, around 20°W/50°N. An area of robust SLP decline also extends over the whole Mediterranean Basin and Northern Africa, and it strengthens the anomalous SLP gradient over Western Europe, thus enhancing the circulation response. Overall, uncertainties in moisture flux projections are linked to the magnitude of this SLP gradient. GCMs projecting an absolute decline in zonal moisture advection (Figure 7.10b) also project a larger-than-average SLP increase from the North Atlantic to Scandinavia, and larger-than-average decrease over continental Europe (Figure 7.12b). Results are opposite for the GCMs in which zonal moisture flux trends are most positive (Figure 7.12c). Interestingly, the difference between these two groups of models is not largest in the area of robust multi-model SLP increase west of the British Isles. The spread in moisture flux trends is therefore related both to the magnitude and the pattern of future JJA SLP changes.

Discussion

Results from the present study confirm the complexity behind summer drying projections over Europe, but also suggest a comprehensive understanding of the physical mechanisms responsible, as shown in Figure 7.13. In keeping with Brogli et al. (2019), our results suggest a prominent role for the land-ocean warming contrast, specifically the reduced North Atlantic warming, although primarily through its impact on regional circulation rather than through thermodynamics exclusively. Indeed, like Haarsma et al. (2015), we find a critical influence of North Atlantic SLP trends on Northern European precipitation, which we interpret through their impact on incoming moisture fluxes into Europe. However, we argue also for an important effect from the Mediterranean heat low that results from enhanced surface warming in Southern Europe. Yet, unlike Seager et al. (2014), and in agreement with Boé et al. (2009), we find that the influence of summer dynamics is largely confined northwards of 45°N. South of that latitude, soil moisture-precipitation feedbacks, triggered by winter and spring drying, are dominant (Rowell and Jones 2006; Boé and Terray 2014).

These mechanisms are also consistent with analyses of inter-annual variability in the summer storm track over Northern Europe (Dong et al. 2013) as well as of spring precipitation and its link to summer temperatures (Vautard et al. 2007; Wang et al. 2011). Still, additional research is needed to quantify the respective importance and magnitude of the various physical mechanisms causing the summer drying. Such an analysis should rely on a common modeling framework with carefully designed experiments to investigate the effects of winter/spring precipitation anomalies or summer circulation changes. It would be critical in order to determine the effects of uncertainties in the magnitude of these physical drivers on future summer climate trends.

Mechanisms of spring circulation change

On its own, the dynamical adjustment approach we adopted only highlights the consistency between large-scale, upper-tropospheric circulation changes and SLP trends over the Mediterranean. Assuming the future upper-tropospheric pattern as given, the near-surface response (winds and SLP) can be understood through the prism of the external mode, the dominant mode of variability in the vertical structure of the seasonal-mean circulation in mid-latitudes (Chapter 3). The enhanced near-surface response over the Mediterranean is caused by low vertical shear, as this region lies in the exit region of the North Atlantic storm track. Yet, the question remains as to what mechanisms are responsible for future upper-tropospheric circulation trends. Their similarity to winter changes suggests that mechanisms identified for winter still hold, i.e., a lengthening of the stationary wavelengths for planetary waves following the strengthening of the subtropical jet (Simpson et al. 2016). However, changes in e.g., the stratospheric vortex could also modulate the response (Zappa 2019). A detailed analysis of inter-model uncertainty in either winter or spring upper-tropospheric circulation trends is also still lacking (Sandler and Harnik 2020). The Mediterranean-scale circulation response to SST trends may also feed back onto the large-scale circulation, contributing to the trends projected by the GCMs. We cannot capture this effect in RCM simulations, since boundary conditions cannot be affected by perturbations within the domain. A global climate model would be required to investigate this point further.

Soil moisture-precipitation feedbacks

The notable sensitivity of summer climate in the southern half of the European domain to cold season P-E is consistent with previous results. Transitional regions between arid and wet climates are generally known to be hot spots of land-atmosphere coupling (Seneviratne et al. 2010; Orth and Seneviratne 2017). At the European scale, the analysis by Knist et al. (2017) based on the EURO- CORDEX model ensemble also confirmed that southern Europe was characterized by strong land atmosphere coupling, unlike northern Europe. Evidence was found that inter-annual variability in winter and spring soil moisture in southern Europe impacted summer precipitation (Vautard et al. 2007) and the likelihood of summer hot days (Quesada et al. 2012; Mueller and Seneviratne 2012). For precipitation, the link may emerge not so much from the direct input of moisture to the atmosphere by evaporation from wet soils as from the impact that higher absolute humidity has on lower troposphere stability (Eltahir 1998; Seneviratne et al. 2010). Downstream precipitation of evaporated soil moisture may still matter locally, however. Soil moisture-precipitation feedbacks thus offer a physically-based causal interpretation of the correlation results from Figures 7.3 and 7.4.

This causality could potentially be called into question if cold-season P-E and summer climate were both impacted by a third, independent process, for instance a persistence in atmospheric dynamics or SSTs that would lead to larger-than-average precipitation from winter to summer. There is however no clear evidence for such a process. The large contribution of surface evaporation to the summertime European moisture budget supports the idea that the correlation is linked to soil moisture-precipitation feedbacks. Simulations in which spring and summer soil moisture are set to a baseline value (and thus decoupled from cold-season P-E variability) could help shed more light on this point.

Results suggest that land-atmosphere coupling also explains much of the projected summer P decline in the southern part of our domain. There, the projected summer decline in E and soil moisture, and the weak P-E trends (Figure 7.5), are consistent with soil moisture-precipitation feedbacks being responsible for the concurrent P decline. Summer E notably decreases most over the Iberian Peninsula and the Balkans, the regions of our domain that experience the most severe P, P-E and soil moisture declines during the cold season. These are also regions with notable land-atmosphere coupling in the present-day climate, i.e., where summer E and low-level RH are most constrained by late spring soil moisture and winter/spring P-E (Figures 7.3e, 4). Notably, over the Iberian Peninsula (and Northwestern Africa, outside of our domain), E is even projected to decline during MAM. This suggests that soil moisture constraints on E will in the future already be felt during spring, and that soil moisture depletion will occur earlier in the year.

By contrast, the fact that, over the northern half of the domain, summer E remains constant in future projections (Figure 7.5f), despite lower P and soil moisture, is evidence that soil moisture feedbacks play a limited role there. Water saturation at the surface decreases, but likely remains high enough, combined with the enhanced evaporative demand in a warmer and drier climate, to maintain E at its historical value. The P-E decline is driven by the decline in P. Future summer P-E becomes clearly negative (as opposed to a historical value close to 0, Figure 7.3a), a change sustained by enhanced depletion of soil moisture. This suggests that land-atmosphere coupling will strengthen over the northern half of the domain, as summer climate becomes increasingly reliant on cold-season accumulated soil moisture.

Still, it should be noted that CMIP5 GCMs may overestimate the strength of soil moisture precipitation feedbacks due to their coarse resolution (Hohenegger et al. 2009). Biases in cloud parametrization or in the effect of anthropogenic aerosols in GCMs could also matter, and explain in particular why regional simulations may exhibit weaker warming and drying trends than their driving GCMs (Boé et al. 2020a).

Future trends in summer circulation

The section discussing moisture flux changes showed that future summer circulation trends strongly constrained changes in zonal moisture advection, thus impacting the future European-domain moisture budget. First, positive SLP anomalies over the North Atlantic are thought to result from the weak SST warming trends in the region, also known as North Atlantic Warming Hole

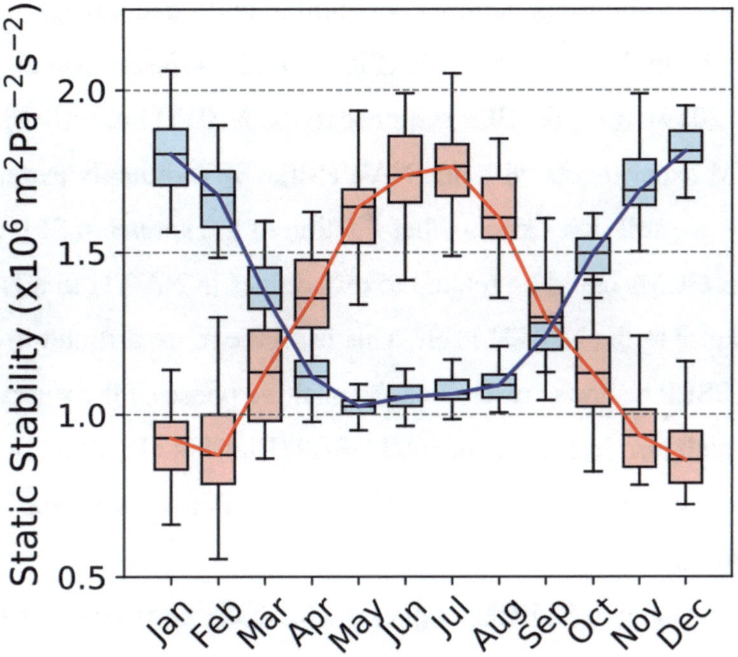

Figure 7.14: Annual cycles of static stability $\sigma = \frac{RT}{p}\frac{d\,log(\theta)}{dp}$ *over land (10°W-30°E/30°-50°N; blue) and the Mediterranean Sea (red) in ERA5, averaged over the 1979-2019 period. Whiskers indicate the full range of inter-annual variability.*

(NAWH). The NAWH believed to follow from a slowdown of the Atlantic Meridional Overturning Circulation, although other forcings are not to be excluded (Hu and Fedorov 2020). Numerical simulations in which global climate models were forced with negative SST anomalies mimicking the NAWH, exhibited an SLP response generally consistent in pattern and magnitude with the projected CMIP5 multi-model average (Figure 7.12a) (Haarsma et al. 2015; Gervais et al. 2019). Still, the SLP response to the NAWH has only been analyzed in single-GCM experiments with one NAWH-like SST anomaly pattern only. Consequently, we still don't know what fraction of the spread in SLP trends across CMIP5 GCMs could be related to differences in NAWH trends themselves, or in the response to the NAWH itself. This is all the more difficult to determine as the NAWH SLP response resembles the positive phase of the summer NAO, changes in which also influence mean JJA SLP (Bladé et al. 2012). In particular, it remains unclear why some GCMs would project SLP increases over Scandinavia (Figure 7.12b).

Second, negative SLP trends over the Mediterranean region have been shown to result from a heat low phenomenon: as the intense surface warming, caused by the decrease in summer precipitation, extends into the lower troposphere, it triggers an overall cyclonic anomaly around the area (Haarsma et al. 2009). This is in sharp contrast with winter and spring trends (Figures 7.6a and 7.7a). Indeed, land surface warming around the Mediterranean is by far largest during summer (Lionello and Scarascia 2018). In addition, near-surface static stability over land is lowest during summer (Figure 7.14), thus favoring the upward propagation of the surface warming and the associated circulation response. By contrast, in winter, near-surface static stability is large over land, but

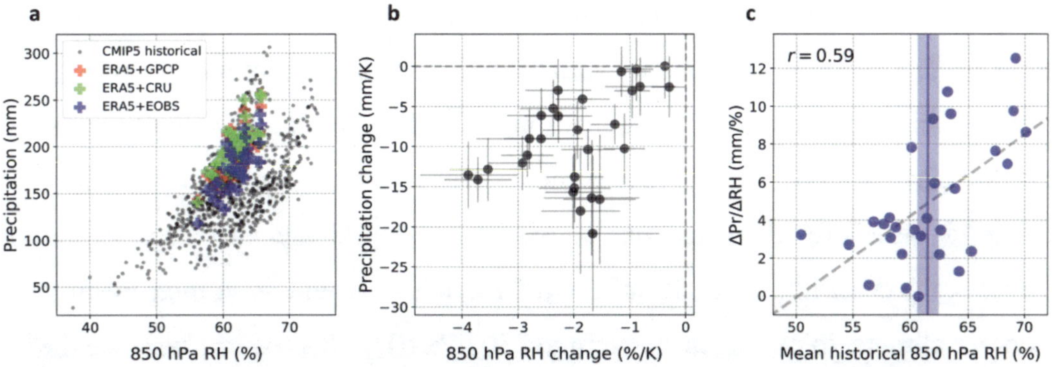

Figure 7.15: (a) Inter-annual variability in JJA European (10°W-30°E/30-50°N) 850 hPa RH and precipitation: output from CMIP5 models (1976-2005, black), and data from ERA5 (for RH) and GPCP (red), CRU TS4 (green) and EOBS (blue) (for precipitation). (b) Change in JJA European precipitation against change in 850 hPa RH across CMIP5 models. 95% confidence intervals are shown by light gray lines. (c) Sensitivity of projected JJA precipitation change to RH change across CMIP5 models, against mean historical JJA European RH. The ERA5 mean summer RH is shown by the thick vertical blue line, along with its 95% confidence interval.

low over the Mediterranean Sea. This explains why the relative SST cooling spreads upwards into the lower troposphere during winter only, generating an anomalous SLP ridge (Chapter 3).

Summer RH and precipitation

Our focus here has been primarily on moisture fluxes and near-surface RH, instead of precipitation. This is in part physically motivated, e.g., since dynamics act on precipitation through their effect on the moisture budget, but also because CMIP5 models are known to differ widely in their representation of feedbacks between atmospheric humidity, clouds and surface processes. Indeed, differences in the representation of land-atmosphere coupling, notably cloud feedbacks, were found to be related to uncertainties in projections of summer precipitation across global and regional climate models (Boé and Terray 2014), with evidence pointing towards noticeable model deficiencies likely impacting future projections (Boberg and Christensen 2012; Orth et al. 2016; Boé et al. 2020a,b). This added uncertainty might therefore complicate the analysis of inter-model scatter. We show on Figure 7.15a inter-annual variability in summer precipitation and 850 hPa RH across observational, reanalysis and CMIP5 data. It is striking to observe that the CMIP5 models neatly divide themselves into two groups with similar precipitation-RH relationships; one consisent with ERA5 and various precipitation datasets, the other not. In addition, CMIP5 models vary widely in their mean summer climate, in particular in average 850 hPa RH. This matters because the precipitation-RH relationship is not linear, precipitation being less sensitive to RH variability as mean RH decreases (Figure 7.15a). This leads to poor correlation between JJA RH and precipitant changes in CMIP5 projections (Figure 7.15b). The non-uniqueness of the relationship stands in contrast to other seasons, when the share of convective precipitation is lower, and dynamics matter more than model parametrizations. Consistent with Figure 7.15a, mean historical RH provides a useful constraint on the sensitivity of mean precipitation to changes in background RH (Figure 7.15c). ERA5 indicates a mean RH value of 61-62% over Europe during summer, which falls roughly along the CMIP5 median. This adds further uncertainty to future precipitation trends, since for same RH change, different models may project widely different P changes.

In particular, some models may be less reliable for precipitation projections because they are too sensitive to variations in low-level RH.

Conclusion

The broad division of continental Europe into a southern region, dependent on evaporation of cold-season water storage to maintain summer atmospheric humidity, and a northern region reliant on the narrow North Atlantic moisture jet, emerges from the analysis of historical summertime moisture fluxes. This dichotomy, though imperfect, is useful to understand future summer projections over the European domain. Since evapotranspiration is a primary source of moisture south of 45°N, the region is logically highly sensitive to variations in the winter/spring moisture budget. It is led towards robust drying, dictated by soil moisture-precipitation feedbacks, as cold-season precipitation declines, due to changes in regional circulation caused by shifts in planetary waves and a relative cooling of Mediterranean SSTs compared to land. Those feedbacks also lead to enhanced summer warming and to the development of a Mediterranean-wide heat low that contributes to weakening the incoming moisture transport towards Northwestern Europe upon which this last region depends on to maintain its moisture during summer. In parallel, a large-scale land/ocean warming contrast over the North Atlantic, enhanced by the weakening of the AMOC, is followed by the development of anomalous anticyclonic circulation west of the British Isles, which also acts to reduce the westerly transport of moisture towards the continent.

While some of the various physical mechanisms presented here have been previously discussed, this study has sought to provide a comprehensive, physical picture of summer European drying, and to describe how these different mechanisms interact with each other. We also anchored summer climate change over Southern Europe into a physical understanding of cold-season precipitation trends. Several major sources of uncertainty were discussed, including moisture flux projections and the sensitivity of regional precipitation to RH. The adoption of a common modeling framework may help to better assess the respective role of these various mechanisms in future summer climate trends.

Acknowledgments. We acknowledge the World Climate Research Programme's Working Group on Coupled Modelling, which is responsible for CMIP, and we thank the climate modeling groups for producing and making available their model output. This work was funded by the Office Chérifien des Phosphates (OCP) through Université Mohamed VI Polytechnique, Morocco.

A version of this article first appeared as:
Tuel, A. and Eltahir, E., 2021. Mechanisms of European summer drying under climate change. Journal of Climate, pp.1-51. (Original article under AMS copyright)

References

Chapter 2

[1] Adler, R. F., and Coauthors, (2003): The Version-2 Global Precipitation Climatology Project (GPCP) monthly precipitation analysis (1979–present). J. Hydrometeor., 4, 1147–1167, https://doi.org/ 10.1175/1525-7541(2003)004,1147:TVGPCP.2.0.CO;2.

[2] Allan, R. J., and T. J. Ansell, (2006): A new globally complete monthly historical mean sea level pressure dataset (HadSLP2): 1850–2004. J. Climate, 19, 5816–5842, https://doi.org/10.1175/JCLI3937.1.

[3] Barnes, E. A., and L. Polvani, (2013): Response of the midlatitude jets, and of their variability, to increased greenhouse gases in the CMIP5 models. J. Climate, 26, 7117–7135, https://doi.org/ 10.1175/JCLI-D-12-00536.1.

[4] Bolle, H.-J., (2003): Climate, climate variability, and impacts in the Mediterranean area: An overview. Mediterranean Climate: Variability and Trends, H.-J. Bolle, Ed., Springer, 5–86, https:// doi.org/10.1007/978-3-642-55657-9_2.

[5] Brandefelt, J., and H. Krnich, (2008): Northern Hemisphere sta- tionary waves in future climate projections. J. Climate, 21, 6341–6353, https://doi.org/10.1175/2008JCLI2373.1.

[6] Brogli, R., S. L. Sørland, N. Kröner, and C. Schär, (2019): Causes of future Mediterranean precipitation decline depend on the season. Environ. Res. Lett., 14, 114017, https://doi.org/10.1088/ 1748-9326/ab4438.

[7] Byrne, M. P., and P. O'Gorman, (2016): Understanding decreases in land relative humidity with global warming: Conceptual model and GCM simulations. J. Climate, 29, 9045–9061, https://doi.org/10.1175/JCLI-D-16-0351.1.

[8] Byrne, M. P., and P. O'Gorman, (2018): Trends in continental temperature and hu- midity directly linked to ocean warming. Proc. Natl. Acad. Sci. USA, 115, 4863–4868, https://doi.org/10.1073/pnas.1722312115. Compo, G. P., and Coauthors, 2011: The Twentieth Century Reanalysis project. Quart. J. Roy. Meteor. Soc., 137, 1–28, https://doi.org/10.1002/qj.776.

[9] Dee, D. P., and Coauthors, (2011): The ERA-Interim reanalysis: Configuration and performance of the data assimilation sys- tem. Quart. J. Roy. Meteor. Soc., 137, 553–597, https://doi.org/ 10.1002/qj.828.

[10] Deser, C., L. Terray, and A. S. Phillips, (2016): Forced and internal components of winter air temperature trends over North America during the past 50 years: Mechanisms and implications. J. Climate, 29, 2237–2258, https://doi.org/10.1175/JCLI-D-15-0304.1.

[11] Gervais, M., J. Shaman, and Y. Kushnir, (2019): Impacts of the North Atlantic warming hole in future climate projections: Mean atmospheric circulation and the North Atlantic jet. J. Climate, 32, 2673–2689, https://doi.org/10.1175/JCLI-D-18-0647.1.

[12] Gianotti, R. L., and E. A. B. Eltahir, (2014a): Regional climate mod- eling over the Maritime Continent. Part I: New parameteriza- tion for convective cloud fraction. J. Climate, 27, 1488–1503, https://doi.org/10.1175/JCLI-D-13-00127.1.

[13] Gianotti, R. L., and E. A. B. Eltahir, (2014b): Regional climate modeling over the Maritime Continent. Part II: New parameterization for auto- conversion of convective rainfall. J. Climate, 27, 1504–1523, https://doi.org/10.1175/JCLI-D-13-00171.1.

[14] Gianotti, R. L., D. Zhang, and E. A. B. Eltahir, (2012): Assessment of the Regional

Climate Model version 3 over the Maritime Continent using dif- ferent cumulus parameterization and land surface schemes. J. Climate, 25, 638–656, https://doi.org/10.1175/JCLI-D-11-00025.1.

[15] Giorgi, F., and X. Bi, (2005): Updated regional precipitation and temperature changes for the 21st century from ensembles of recent AOGCM simulations. Geophys. Res. Lett., 32, L21715, https://doi.org/10.1029/2005GL024288.

[16] Giorgi, F., and P. Lionello, (2008): Climate change projections for the Mediterranean region. Global Planet. Change, 63, 90–104, https://doi.org/10.1016/j.gloplacha.2007.09.005.

[17] Hall, N. M., H. Lin, and J. Derome, (2001): The extratropical signal generated by a midlatitude SST anomaly. Part II: Influence on seasonal forecasts. J. Climate, 14, 2696–2709, https://doi.org/ 10.1175/1520-0442(2001)014,2696:TESGBA.2.0.CO;2.

[18] Harris, I., P. D. Jones, T. J. Osborn, and D. H. Lister, (2014): Updated high-resolution grids of monthly climatic observations—The CRU TS3.10 Dataset. Int. J. Climatol., 34, 623–642, https:// doi.org/10.1002/joc.3711.

[19] He, J., and B. J. Soden, (2017): A re-examination of the projected subtropical precipitation decline. Nat. Climate Change, 7, 53–57, https://doi.org/10.1038/nclimate3157.

[20] Held, I. M., M. Ting, and H. Wang, (2002): Northern winter stationary waves: Theory and modeling. J. Climate, 15, 2125–2144, https:// doi.org/10.1175/1520-0442(2002)015,2125:NWSWTA.2.0.CO;2.

[21] Hoerling, M., J. Eischeid, J. Perlwitz, X. Quan, T. Zhang, and P. Pegion, (2012): On the increased frequency of Mediterranean drought. J. Climate, 25, 2146–2161, https://doi.org/10.1175/ JCLI-D-11-00296.1.

[22] Hoskins, B. J., and D. J. Karoly, (1981): The steady linear re- sponse of a spherical atmosphere to thermal and orographic forcing. J. Atmos. Sci., 38, 1179–1196, https://doi.org/10.1175/ 1520-0469(1981)038,1179:TSLROA.2.0.CO;2.

[23] Huang, B., and Coauthors, (2017): Extended Reconstructed Sea Surface Temperature, version 5 (ERSSTv5): Upgrades, vali- dations, and intercomparisons. J. Climate, 30, 8179–8205, https://doi.org/10.1175/JCLI-D-16-0836.1.

[24] Hurrell, J. W., (1995): Decadal trends in the North Atlantic oscilla- tion: Regional temperatures and precipitation. Science, 269, 676–679, https://doi.org/10.1126/science.269.5224.676.

[25] Im, E.-S., and E. A. B. Eltahir, (2018a): Simulation of the diurnal variation of rainfall over the western Maritime Continent using a regional climate model. Climate Dyn., 51, 73–88, https://doi.org/10.1007/s00382-017-3907-3.

[26] Im, E.-S., and E. A. B. Eltahir, (2018b): Simulations of the observed 'jump' in the West African monsoon and its underlying dynamics using the MIT regional climate model. Int. J. Climatol., 38, 841–852, https://doi.org/10.1002/joc.5214.

[27] Kalnay, E., and Coauthors, (1996): The NCEP/NCAR 40-Year Reanalysis Project. Bull. Amer. Meteor. Soc., 77, 437–472, https:// doi.org/10.1175/1520-0477(1996)077,0437:TNYRP.2.0.CO;2.

[28] Kelley, C., M. Ting, R. Seager, and Y. Kushnir, (2012): The relative contributions of radiative forcing and internal climate variabil- ity to the late 20th century winter drying of the Mediterranean region. Climate Dyn., 38, 2001–2015, https://doi.org/10.1007/ s00382-

011-1221-z.
[29] Kelley, C., S. Mohtadi, M. A. Cane, R. Seager, and Y. Kushnir, (2015): Climate change in the Fertile Crescent and implications of the recent Syrian drought. Proc. Natl. Acad. Sci. USA, 112, 3241– 3246, https://doi.org/10.1073/pnas.1421533112.
[30] Lu, J., G. A. Vecchi, and T. Reichler, (2007): Expansion of the Hadley cell under global warming. Geophys. Res. Lett., 34, L06805, https://doi.org/10.1029/2006GL028443.
[31] Pal, J. S., and E. A. B. Eltahir, (2016): Future temperature in southwest Asia projected to exceed a threshold for human adaptability. Nat. Climate Change, 6, 197–200, https://doi.org/10.1038/nclimate2833.
[32] Pal, J. S., and Coauthors, (2007): Regional climate modeling for the developing world: The ICTP RegCM3 and RegCNET. Bull. Amer. Meteor. Soc., 88, 1395–1410, https://doi.org/10.1175/BAMS-88-9-1395.
[33] Planton, S., and Coauthors, (2012): The climate of the Mediterranean region in future climate projections. The Climate of the Mediterranean Region, P. Lionello, Ed., Elsevier, 449–502, https://doi.org/10.1016/B978-0-12-416042-2.00008-2.
[34] Poli, P., and Coauthors, (2016): ERA-20C: An atmospheric re- analysis of the twentieth century. J. Climate, 29, 4083–4097, https://doi.org/10.1175/JCLI-D-15-0556.1.
[35] Rayner, N. A., D. E. Parker, E. B. Horton, C. K. Folland, L. V. Alexander, D. P. Rowell, E. C. Kent, and A. Kaplan, (2003): Global analyses of sea surface temperature, sea ice, and night marine air temperature since the late nineteenth century. J. Geophys. Res., 108, 4407, https://doi.org/10.1029/2002JD002670.
[36] Reynolds, R. W., T. M. Smith, C. Liu, D. B. Chelton, K. S. Casey, and M. G. Schlax, (2007): Daily high-resolution blended ana- lyses for sea surface temperature. J. Climate, 20, 5473–5496, https://doi.org/10.1175/2007JCLI1824.1.
[37] Rodwell, M. J., and B. J. Hoskins, (1996): Monsoons and the dy- namics of deserts. Quart. J. Roy. Meteor. Soc., 122, 1385–1404, https://doi.org/10.1002/qj.49712253408.
[38] Rojas, M., L. Z. Li, M. Kanakidou, N. Hatzianastassiou, G. Seze, and H. L. Treut, (2013): Winter weather regimes over the Mediterranean region: Their role for the regional climate and projected changes in the twenty-first century. Climate Dyn., 41, 551–571, https://doi.org/10.1007/s00382-013-1823-8.
[39] Sannino, G., A. Bargagli, and V. Artale, (2002): Numerical modeling of the mean exchange through the Strait of Gibraltar. J. Geophys. Res., 107, 3094, https://doi.org/10.1029/2001JC000929.
[40] Scheff, J., and D. Frierson, (2012): Twenty-first-century multi- model subtropical precipitation declines are mostly midlat- itude shifts. J. Climate, 25, 4330–4347, https://doi.org/10.1175/ JCLI-D-11-00393.1.
[41] Seager, R., H. Liu, N. Henderson, I. R. Simpson, C. Kelley, T. Shaw, Y. Kushnir, and M. Ting, (2014): Causes of increasing aridification of the Mediterranean region in response to rising greenhouse gases. J. Climate, 27, 4655–4676, https://doi.org/ 10.1175/JCLI-D-13-00446.1.
[42] Seager, R., T. Osborn, Y. Kushnir, I. Simpson, J. Nakamura, and H. Liu, (2019): Climate variability and change of Mediterranean-type climates. J. Climate, 32, 2887–2915, https://doi.org/10.1175/ JCLI-D-18-0472.1.
[43] Simpson, I. R., R. Seager, M. Ting, and T. A. Shaw, (2016): Causes of change in Northern Hemisphere winter meridional winds and regional hydroclimate. Nat. Climate

Change, 6, 65–70, https:// doi.org/10.1038/nclimate2783.
[44] Simpson, I. R., P. Hitchcock, R. Seager, Y. Wu, and P. Callaghan, (2018): The downward influence of uncertainty in the Northern Hemisphere stratospheric polar vortex response to climate change. J. Climate, 31, 6371–6391, https://doi.org/10.1175/ JCLI-D-18-0041.1.
[45] Sutton, R. T., B. Dong, and J. M. Gregory, (2007): Land/sea warming ratio in response to climate change: IPCC AR4 model results and comparison with observations. Geophys. Res. Lett., 34, L02701, https://doi.org/10.1029/2006GL028164.
[46] Taylor, K. E., R. J. Stouffer, and G. A. Meehl, (2012): An overview of CMIP5 and the experiment design. Bull. Amer. Meteor. Soc., 93, 485–498, https://doi.org/10.1175/BAMS-D-11-00094.1.
[47] Tuel, A., and E. A. B. Eltahir, (2018): Large-scale circulation and precipitation decline in the Mediterranean. 2018 Fall Meeting, Washington, DC, Amer. Geophys. Union, Abstract A41F-08. Vose, R., and Coauthors, 2012: NOAA's Merged Land–Ocean Surface Temperature analysis. Bull. Amer. Meteor. Soc., 93, 1677–1685, https://doi.org/10.1175/BAMS-D-11-00241.1.
[48] Winter, J. M., J. S. Pal, and E. A. B. Eltahir, (2009): Coupling of integrated biosphere simulator to Regional Climate Model version 3. J. Climate, 22, 2743–2757, https://doi.org/10.1175/ 2008JCLI2541.1.
[49] Woollings, T., J. M. Gregory, J. G. Pinto, M. Reyers, and D. J. Brayshaw, (2012): Response of the North Atlantic storm track to climate change shaped by ocean–atmosphere coupling. Nat. Geosci., 5, 313–317, https://doi.org/10.1038/ngeo1438.
[50] Yin, J. H., (2005): A consistent poleward shift of the storm tracks in simulations of 21st century climate. Geophys. Res. Lett., 32, L18701, https://doi.org/10.1029/2005GL023684.
[51] Zanchettin, D., A. Rubino, D. Matei, O. Bothe, and J. Jungclaus, (2013): Multidecadal-to-centennial SST variability in the MPI- ESM simulation ensemble for the last millennium. Climate Dyn., 40, 1301–1318, https://doi.org/10.1007/s00382-012-1361-9.
[52] Zappa, G., and T. G. Shepherd, (2017): Storylines of atmospheric circulation change for European regional climate impact as- sessment. J. Climate, 30, 6561–6577, https://doi.org/10.1175/JCLI-D-16-0807.1.
[53] Zappa, G., M. K. Hawcroft, L. Shaffrey, E. Black, and D. J. Brayshaw, (2015a): Extratropical cyclones and the projected decline of winter Mediterranean precipitation in the CMIP5 models. Climate Dyn., 45, 1727–1738, https://doi.org/10.1007/ s00382-014-2426-8.
[54] Zappa, G., B. J. Hoskins, and T. G. Shepherd, (2015b): The dependence of wintertime Mediterranean precipitation on the atmospheric circulation response to climate change. Environ. Res. Lett., 10, 104012, https://doi.org/10.1088/1748-9326/10/10/104012.

Chapter 3

[1] Blackmon, M. L., R. A. Madden, J. M. Wallace, and D. S. Gutzler, (1979): Geographical variations in the vertical structure of geopotential height fluctuations. J. Atmos. Sci., 36, 2450–2466, https://doi.org/ 10.1175/15200469(1979)036¸2450:GVITVS.2.0.CO;2.
[2] Brogli, R., S. Sørland, N. Kröner, and C. Schär, (2019): Causes of future

Mediterranean precipitation decline depend on the season. Environ. Res. Lett., 14, 114017, https://doi.org/10.1088/ 1748-9326/ab4438.

[3] Byrne, M. P., and P. O'Gorman, (2018): Trends in continental tem- perature and humidity directly linked to ocean warming. Proc. Natl. Acad. Sci. USA, 115, 4863–4868, https://doi.org/10.1073/ pnas.1722312115.

[4] Dee, D. P., and Coauthors, (2011): The ERA-Interim reanalysis: Configuration and performance of the data assimilation sys- tem. Quart. J. Roy. Meteor. Soc., 137, 553–597, https://doi.org/ 10.1002/qj.828.

[5] Gates, W., and Coauthors, (1999): An overview of the results of the Atmospheric Model Intercomparison Project (AMIP I). Bull. Amer. Meteor. Soc., 80, 29–55, https://doi.org/10.1175/1520- 0477(1999)080¸0029:AOOTRO˛2.0.CO;2.

[6] Giorgi, F., and P. Lionello, (2008): Climate change projections for the Mediterranean region. Global Planet. Change, 63, 90–104, https://doi.org/10.1016/j.gloplacha.2007.09.005.

[7] Hall, A., P. Cox, C. Huntingford, and S. Klein, (2019): Progressing emergent constraints on future climate change. Nat. Climate Change, 9, 269–278, https://doi.org/10.1038/s41558-019-0436-6. Harnik, N., E. Galanti, O. Martius, and O. Adam, 2014: The anomalous merging of the African and North Atlantic jet streams during the Northern Hemisphere winter of 2010. J. Climate, 27, 7319–7334, https://doi.org/10.1175/JCLI-D-13-00531.1.

[8] Held, I. M., R. Panetta, and R. Pierrehumbert, (1985): Stationary external Rossby waves in vertical shear. J. Atmos. Sci., 42, 865–883, https:// doi.org/10.1175/1520-0469(1985)042¸0865:SERWIV˛2.0.CO;2.

[9] Held, I. M., M. Ting, and H. Wang, (2002): Northern winter stationary waves: Theory and modeling. J. Climate, 15, 2125–2144, https://doi.org/ 10.1175/1520-0442(2002)015¸2125:NWSWTA˛2.0.CO;2.

[10] Hoskins, B. J., and D. J. Karoly, (1981): The steady linear response of a spherical atmosphere to thermal and orographic forcing. J. Atmos. Sci., 38, 1179–1196, https://doi.org/10.1175/1520- 0469(1981)038¸1179:TSLROA˛2.0.CO;2.

[11] Hoskins, B. J., and T. Ambrizzi, (1993): Rossby wave propagation on a realistic longitudinally varying flow. J. Atmos. Sci., 50, 1661–1671, https:// doi.org/10.1175/1520-0469(1993)050¸1661:RWPOAR˛2.0.CO;2.

[12] Rind, D., (2008): The consequences of not knowing low- and high- latitude climate sensitivity. Bull. Amer. Meteor. Soc., 89, 855– 864, https://doi.org/10.1175/2007BAMS2520.1.

[13] Rodwell, M. J., and B. J. Hoskins, (1996): Monsoons and the dy- namics of deserts. Quart. J. Roy. Meteor. Soc., 122, 1385–1404, https://doi.org/10.1002/qj.49712253408.

[14] Seager, R., H. Liu, N. Henderson, I. R. Simpson, C. Kelley, T. Shaw, Y. Kushnir, and M. Ting, (2014): Causes of increasing aridification of the Mediterranean region in response to rising greenhouse gases. J. Climate, 27, 4655–4676, https://doi.org/ 10.1175/JCLI-D-13-00446.1.

[15] Seager, R., T. Osborn, Y. Kushnir, I. Simpson, J. Nakamura, and H. Liu, (2019): Climate variability and change of Mediterranean-type climates. J. Climate, 32, 2887–2915, https://doi.org/10.1175/JCLI-D-18-0472.1. Simpson, I. R., R. Seager, T. Shaw, and M. Ting, 2015: Mediterranean summer climate and the importance of

Middle East topography. J. Climate, 28, 1977–1996, https://doi.org/ 10.1175/JCLI-D-14-00298.1.
[16] Seager, R., T. Osborn, M. Ting, and T. Shaw, (2016): Causes of change in Northern Hemisphere winter meridional wind and regional hydroclimate. Nat. Climate Change, 6, 65–70, https://doi.org/10.1038/nclimate2783.
[17] Seager, R., and Coauthors, (2020): An evaluation of the large-scale at- mospheric circulation and its variability in CESM2 and other CMIP models. J. Geophys. Res. Atmos., 125, e2020JD032835, https://doi.org/10.1029/2020JD032835.
[18] Sutton, R. T., B. Dong, and J. M. Gregory, (2007): Land/sea warming ratio in response to climate change: IPCC AR4 model results and comparison with observations. Geophys. Res. Lett., 34, L02701, https://doi.org/10.1029/2006GL028164.
[19] Taylor, K. E., R. J. Stouffer, and G. A. Meehl, (2012): An overview of CMIP5 and the experiment design. Bull. Amer. Meteor. Soc., 93, 485–498, https://doi.org/10.1175/BAMS-D-11-00094.1.
[20] Tuel, A., and E. Eltahir, (2020): Why is the Mediterranean a climate change hotspot? J. Climate, 33, 5829–5843, https://doi.org/ 10.1175/JCLI-D-19-0910.1.
[21] Wills, R. C. J., R. H. White, and X. J. Levine, (2019): Northern Hemisphere stationary waves in a changing climate. Curr. Climate Change Rep., 5, 372–389, https://doi.org/10.1007/ s40641-019-00147-6.
[22] Zanchettin, D., A. Rubino, D. Matei, O. Bothe, and J. Jungclaus, (2013): Multidecadal-to-centennial SST variability in the MPI-ESM simulation ensemble for the last millennium. Climate Dyn., 40, 1301–1318, https://doi.org/10.1007/s00382-012-1361-9.
[23] Zappa, G., and T. G. Shepherd, (2017): Storylines of atmospheric circulation change for European regional climate impact as- sessment. J. Climate, 30, 6561–6577, https://doi.org/10.1175/ JCLI-D-16-0807.1.
[24] Zappa, G., B. J. Hoskins, and T. G. Shepherd, (2015): The dependence of wintertime Mediterranean precipitation on the atmospheric circulation response to climate change. Environ. Res. Lett., 10, 104012, https://doi.org/10.1088/1748-9326/10/10/104012.

Chapter 4

[1] Abatzoglou JT, Drobowski SZ, Parks SA, Hegewisch K (2018) TerraClimate, a high-resolution global dataset of monthly climate and climatic water balance from 1958–2015. Sci Data 5:170191
[2] Balaghi R, Jlibene M, Tychon B, Heerens H (2012) Agrometeorological forecasting of cereal crop yields in Morocco. Morocco, National Institute for Agronomic Research (INRA)
[3] Boberg F, Christensen J (2012) Overestimation of Mediterranean summer temperature projections due to model deficiencies. Nature Clim Change 2:433–436
[4] Born K, Fink AH, Paeth H (2008) Dry and wet periods in the northwestern Maghreb for present day and future climate conditions. Meteorol Z 17(5):533–551

[5] Coppola E et al (2020) A first-of-its-kind multi-model convection permitting ensemble for investigating convective phenomena over Europe and the Mediterranean. Clim Dyn 55:3–34
[6] Davini P, Cagnazzo C (2014) On the misinterpretation of the North Atlantic Oscillation in CMIP5 models. Clim Dyn 43:1497–1511
[7] Dee DP et al (2011) The ERA-Interim reanalysis: configuration and performance of the data assimilation system. Q J Roy Meteorol Soc 137:553–597
[8] Diffenbaugh NS, Giorgi F (2012) Climate change hotspots in the CMIP5 global climate model ensemble. Climatic Change 114:813–822
[9] Driouech F, Déqué M, Sánchez-Gómez E (2010) Weather regimes–Moroccan precipitation link in a regional climate change simulation. Global Planet Change 72:1–10
[10] Dubrovsky M, Hayes M, Duce P, Trnka M, Svoboda M, Zara P (2014) Multi-GCM projections of future drought and climate variability indicators for the Mediterranean region. Reg Environ Change 14:1907–1919
[11] Dufresne J et al (2013) Climate change projections using the IPSL-CM5 Earth System Model: from CMIP3 to CMIP5. Clim Dyn 40:2123–2165
[12] Dunne JP et al (2012) GFDLs ESM2 global coupled climate-carbon earth system models. Part I: physical formulation and baseline simulation characteristics. J Clim 25:6646–6665
[13] Ehret U, Zehe E, Wulfmeyer V, Warrach-Sagi K, Liebert J (2012) Should we apply bias correction to global and regional climate model data? Hydrol Earth Syst Sci Discuss 9:5355–5387
[14] Fantini A et al (2018) Assessment of multiple daily precipitation statistics in ERA-Interim driven Med-CORDEX and EURO-CORDEX experiments against high resolution observations. Clim Dyn 51:877–900
[15] Gao X, Pal JS, Giorgi F (2006) Projected changes in mean and extreme precipitation over the Mediterranean region from a high resolution double nested RCM simulation. Geophys Res Lett 33:L03706
[16] Giorgetta MA et al (2013) Climate and carbon cycle changes from 1850 to 2100 in MPI-ESM simulations for the coupled model intercomparison project phase 5. J Adv Model Earth Syst 5:572–597
[17] Giorgi F, Lionello P (2008) Climate change projections for the Mediterranean region. Global Planet Change 63:90–104
[18] Gianotti RL (2012) Convective cloud and rainfall processes over the Maritime continent: simulation and analysis of the diurnal cycle. Ph.D. dissertation, Massachusetts Institute of Technology, 306
[19] Gianotti RL, Eltahir EAB (2014a) Regional climate modeling over the Maritime Continent. Part I: new parameterization for convective cloud fraction. J Climate 27:1488–1503
[20] Gianotti RL, Eltahir EAB (2014b) Regional climate modeling over the Maritime Continent. Part II: new parameterization for autoconversion of convective rainfall. J Climate 27:1504–1523
[21] Holtslag AAM, De Bruijn EIF, Pan HL (1990) A high resolution air mass transformation model for shortrange weather forecasting. Mon Wea Rev 118:1561–1575

[22] Huffman GJ et al (2007) The TRMM multi-satellite precipitation analysis (TMPA): quasi-global, multiyear, combined-sensor precipitation estimates at fine scales. J Hydrometeorol 8(1):38–55

[23] Hurrell JW, Kushnir Y, Ottersen G, Visbeck M (2013) An Overview of the North Atlantic Oscillation, In: JW Hurrell, Y Kushnir, G Ottersen, M. Visbeck (eds). The North Atlantic Oscillation: climatic Significance and Environmental Impact. https://doi.org/10.1029/134GM01

[24] Im ES, Gianotti RL, Eltahir EAB (2014) Improving the simulation of the West African Monsoon using the MIT Regional Climate Model. J Clim 27:2209–2229

[25] IPCC (2013) Climate change 2013: the physical science basis. Contribution of working group I to the fifth assessment report of the intergovernmental panel on climate change. Cambridge University Press, Cambridge, New York, p 1535

[26] Jacob D et al (2014) EURO-CORDEX: new high-resolution climate change projections for European impact research. Reg Environ Change 14:563–578

[27] Kalnay E et al (1996) The NCEP/NCAR 40-Year reanalysis project. Bull Amer Meteor Soc 77:437–472

[28] Kiehl JT et al. (1996) Description of the NCAR Community Climate Model (CCM3). NCAR Tech. Note NCAR/TN-4201STR, 152 pp. https://doi.org/10.5065/D6FF3Q99

[29] Knippertz P, Christoph M, Speth P (2003) Long-term precipitation variability in Morocco and the link to the large-scale circulation in recent and future climates. Meteorol Atmos Phys 83:67–88

[30] Liang X, Kunkel K, Meehl G, Jones R, Wang J (2008) Regional climate models downscaling analysis of general circulation models present climate biases propagation into future change projections. Geophys Res Lett 35:L08709

[31] Li H, Sheffield J, Wood EF (2010) Bias correction of monthly precipitation and temperature fields from Intergovernmental Panel on Climate Change AR4 models using equidistant quantile matching. J Geophys Res 115:D10101

[32] Marchane A, Tramblay Y, Hanich L, Ruelland D, Jarlan LD (2017) Climate change impacts on surface water resources in the Rheraya catchment (high-atlas, Morocco). Hydrol Sci J 62(6):979–995

[33] Marcella M, Eltahir EAB (2012) Modeling the summertime climate of Southwest Asia: the role of land surface processes in shaping the climate of semiarid regions. J Climate 25:704–719

[34] Marcella M, Eltahir EAB (2014) Introducing an irrigation scheme to a regional climate model: a case study over West Africa. J Climate 27:5708–5723

[35] McSweeney CF, Jones RG, Lee RW, Rowell DP (2015) Selecting CMIP5 GCMs for downscaling over multiple regions. Clim Dyn 44:3237–3260

[36] MEMEE (2009) Stratégie Nationale de l'Eau. Ministère de l'Energie, des Mines, de l'Eau et de l'Environnement

[37] Milewski A, Elkadiri R, Durham M (2015) Assessment and comparison of TMPA satellite precipitation products in varying climatic and topographic regimes in Morocco. Remote Sens 7:5697–5717

[38] Pal JS et al (2007) Regional climate modeling for the developing world: the ICTP RegCM3 and RegCNET. Bull Am Meteorol Soc 88:1395–1409

[39] Pal JS, Small EE, Eltahir EAB (2000) Simulation of regional-scale water and energy budgets: representation of subgrid cloud and precipitation processes within RegCM. J Geophys Res 105:29579–29594
[40] Pal JS, Eltahir EAB (2016) Future temperature in southwest Asia projected to exceed a threshold for human adaptability. Nature Clim Change 6:128–129
[41] Pepin N et al (2015) Elevation-dependent warming in mountain regions of the world. Nat Clim Change 5:424–430
[42] Prein AF et al (2015) A review on regional convection-permitting climate modeling: demonstrations, prospects, and challenges. Rev Geophys 53:323–361
[43] Rangwala I, Miller JR (2012) Climate change in mountains: a review of elevation-dependent warming and its possible causes. Clim Change 114:527–547
[44] Reynolds RW, Rayner NA, Smith TM, Stokes DM, Wang W (2002) An improved in situ and satellite SST analysis for climate. J Clim 15:1609–1625
[45] Romera R, Sánchez E, Domínguez M, Gaertner MA, Gallardo C (2015) Evaluation of present-climate precipitation in 25 km resolution regional climate model simulations over Northwest Africa. Climate Res 66(2):125–139
[46] Seager R, Liu H, Henderson N, Simpson I, Kelley C, Shaw T, Kushnir Y, Ting M (2014) Causes of increasing aridification of the Mediterranean region in response to rising greenhouse gases. J Climate 27:4655–4676
[47] Solman S, Nuñez M, Cabre MF (2008) Regional climate change experiments over southern South America. I: present climate. Clim Dyn 30:533–552
[48] Stéfanon M et al (2014) Soil moisture-temperature feedbacks at meso-scale during summer heat waves over Western Europe. Clim Dyn 42:1309–1324
[49] Taylor KE, Stouffer RJ, Meehl GA (2012) An overview of CMIP5 and the experiment design. Bull Am Meteorol Soc 93:485–498
[50] Tebaldi C, Smith RW, Nychka D, Mearns LO (2005) Quantifying uncertainty in projections of regional climate change: a Bayesian approach to the analysis of multi-model ensembles. J Clim 18:1524–1540
[51] Tramblay Y, Ruelland D, Somot S, Bouchaia R, Servat E (2013) High-resolution Med-CORDEX regional climate model simulations for hydrological impact studies: a first evaluation of the ALADIN-Climate model in Morocco. Hydrol Earth Syst Sci 17:3721–3739
[52] Tramblay Y, Jarlan L, Hanich L, Somot S (2018) Future scenarios of surface water resources availability in north African dams. Water Resour Manage 32:1291–1306
[53] Tuel A, Eltahir EAB (2020) Why is the Mediterranean a Climate Change Hotspot? Accepted in Journal of Climate
[54] Van Vuuren DP et al (2011) The representative concentration pathways: an overview. Climatic Change 109:5–31
[55] Walton DB, Hall A, Berg N, Schwartz M, Sun F (2017) Incorporating snow albedo feedback into downscaled temperature and snow cover projections for California's sierra Nevada. J Climate 30:1417–1438
[56] Wang X, Li J, Sun C, Liu T (2017) NAO and its relationship with the Northern Hemisphere mean surface temperature in CMIP5 simulations. J Geophys Res Atmos 122:4202–4227

[57] Winter JM, Pal JS, Eltahir EAB (2009) Coupling of integrated biosphere simulator to regional climate model version 3. J Clim 22:2743–2757

[58] Zappa G, Hoskins BJ, Shepherd TG (2015) The dependence of wintertime Mediterranean precipitation on the atmospheric circulation response to climate change. Environ Res Lett 10:104012

[59] Zeng X, Zhao M, Dickinson RE (1998) Intercomparison of bulk aerodynamic algorithms for the computation of sea surface fluxes using TOGA COARE and TAO data. J Climate 11:2628–2644

Chapter 5

[1] Anderson, E. A., (2006): Snow Accumulation and Ablation Model - SNOW-17, Tech. Rep. User Documentation, U.S. National Weather Service, Silver Springs, MD.

[2] Berghuijs, W. R., Woods, R. A., and Hrachowitz, M., (2014): A precipitation shift from snow towards rain leads to a decrease in streamflow, Nature Climate Change, 4, 583–586, https://doi.org/10.1038/nclimate2246, http://www.nature.com/articles/nclimate2246.

[3] Born, K., Fink, A. H., and Paeth, H., (2008): Dry and wet periods in the northwestern Maghreb for present day and future climate conditions, Meteorologische Zeitschrift, 17, 533–551, https://doi.org/10.1127/0941-2948/2008/0313.

[4] Boudhar, A., Hanich, L., Boulet, G., Duchemin, B., Berjamy, B., and Chehbouni, A., (2009): Evaluation of the Snowmelt Runoff model in the Moroccan High Atlas Mountains using two snow-cover estimates, Hydrological Sciences Journal, 54, 1094–1113, https://doi.org/10.1623/hysj.54.6.1094.

[5] Boudhar, A., Duchemin, B., Hanich, L., Jarlan, L., Chaponni re, A., Maisongrande, P., Boulet, G., and Chehbouni, A., (2010): Long-term analysis of snow-covered area in the Moroccan High-Atlas through remote sensing, International Journal of Applied Earth Observation and Geoinformation, 12, https://doi.org/10.1016/j.jag.2009.09.008.

[6] Boudhar, A., Boulet, G., Hanich, L., Sicart, J. E., and Chehbouni, A., (2016): Energy fluxes and melt rate of a seasonal snow cover in the Moroccan High Atlas, Hydrological Sciences Journal, 61, 931–943, https://doi.org/10.1080/02626667.2014.965173.

[7] Davenport, F. V., Herrera-Estrada, J. E., Burke, M., and Diffenbaugh, N. S., (2020): Flood Size Increases Nonlinearly Across the Western United States in Response to Lower Snow-Precipitation Ratios, Water Resources Research, 56, https://doi.org/10.1029/2019WR025571, https://onlinelibrary.wiley.com/doi/abs/10.1029/2019WR025571.

[8] Dee, D. P., Uppala, S. M., Simmons, A. J., Berrisford, P., Poli, P., Kobayashi, S., Andrae, U., Balmaseda, M. A., Balsamo, G., Bauer, P., Bechtold, P., Beljaars, A. C., van de Berg, L., Bidlot, J., Bormann, N., Delsol, C., Dragani, R., Fuentes, M., Geer, A. J., Haimberger, L., Healy, S. B., Hersbach, H., H lm, E. V., Isaksen, L., K llberg, P., K hler, M., Matricardi, M., Mcnally, A. P., Monge-Sanz, B. M., Morcrette, J. J., Park, B. K., Peubey, C., de Rosnay, P., Tavolato, C., Th paut, J. N., and Vitart, F., (2011): The ERA-Interim reanalysis: Configuration and performance of the data assimilation system, Quarterly Journal of the Royal Meteorological Society, 137, 553–597, https://doi.org/10.1002/qj.828, https://doi.org/10.1002/qj.828.

[9] Derin, Y., Anagnostou, E., Berne, A., Borga, M., Boudevillain, B., Buytaert, W., Chang, C. H., Delrieu, G., Hong, Y., Hsu, Y. C., Lavado-Casimiro, W., Manz, B., Moges, S.,

Nikolopoulos, E. I., Sahlu, D., Salerno, F., Rodriguez-Sanchez, J. P., Vergara, H. J., and Yilmaz, K. K. (2016): Multiregional satellite precipitation products evaluation over complex terrain, Journal of Hydrometeorology, 17, 1817–1836, https://doi.org/10.1175/JHM-D-15-0197.1.

[10] Duan, K., Sun, G., McNulty, S. G., Caldwell, P. V., Cohen, E. C., Sun, S., Aldridge, H. D., Zhou, D., Zhang, L., and Zhang, Y. (2017): Future shift of the relative roles of precipitation and temperature in controlling annual runoff in the conterminous United States, Hydrology and Earth System Sciences, 21, 5517–5529, https://doi.org/10.5194/hess-21-5517-2017, https://hess.copernicus.org/articles/21/5517/2017/.

[11] El Khalki, E. M., Tramblay, Y., El Mehdi Saidi, M., Bouvier, C., Hanich, L., Benrhanem, M., and Alaouri, M. (2018): Comparison ofmodeling approaches for flood forecasting in the High Atlas Mountains of Morocco, Arabian Journal of Geosciences, 11, 410, https://doi.org/10.1007/s12517-018-3752-7, http://link.springer.com/10.1007/s12517-018-3752-7.

[12] El Moçayd, N., Kang, S., and Eltahir, E. A. (2020): Climate change impacts on theWater Highway project in Morocco, Hydrology and Earth System Sciences, 24, 1467–1483, https://doi.org/10.5194/hess-24-1467-2020, https://hess.copernicus.org/articles/24/1467/2020/.

[13] Follum, M. L., Downer, C. W., Niemann, J. D., Roylance, S. M., and Vuyovich, C. M. (2015): A radiation-derived temperature-index snow routine for the GSSHA hydrologic model, Journal of Hydrology, 529, 723–736, https://doi.org/10.1016/j.jhydrol.2015.08.044.

[14] Hall, D. K. and Riggs, G. A. (2016): MODIS/Terra Snow Cover Daily L3 Global 500m SIN Grid, Version 6. MOD10A1 and MYD10A1, https://doi.org/10.5067/MODIS/MOD10A1.006, https://nsidc.org/data/MOD10A1/versions/6.

[15] Hashemi, H., 375 Nordin, M., Lakshmi, V., Huffman, G. J., and Knight, R. (2017): Bias correction of long-term satellite monthly precipitation product (TRMM 3B43) over the conterminous United States, Journal of Hydrometeorology, 18, 2491–2509, https://doi.org/10.1175/JHM-D-17-0025.1.

[16] Hssaisoune, M., Bouchaou, L., Sifeddine, A., Bouimetarhan, I., and Chehbouni, A. (2020): Moroccan groundwater resources and evolution with global climate changes, Geosciences (Switzerland), 10, 81, https://doi.org/10.3390/geosciences10020081, https://www.mdpi.com/380 2076-3263/10/2/81.

[17] Huffman, G. J., Adler, R. F., Bolvin, D. T., Gu, G., Nelkin, E. J., Bowman, K. P., Hong, Y., Stocker, E. F., and Wolff, D. B. (2007): The TRMM Multisatellite Precipitation Analysis (TMPA): Quasi-global, multiyear, combined-sensor precipitation estimates at fine scales, Journal of Hydrometeorology, 8, 38–55, https://doi.org/10.1175/JHM560.1.

[18] Jarvis, A., Nelson, A., and Guevara, E. (2008): Jarvis, A., H.I. Reuter, A. Nelson, E. Guevara, 2008, Hole-filled SRTM for the globe Version 4, available from the CGIAR-CSI SRTM 90m Database (http://srtm.csi.cgiar.org), http://srtm.csi.cgiar.org.

[19] Knippertz, P., Christoph, M., and Speth, P. (2003): Long-term precipitation variability in Morocco and the link to the large-scale circulation in recent and future climates,

Meteorology and Atmospheric Physics, 83, 67–88, https://doi.org/10.1007/s00703-002-0561-y.

[20] Lopez-Moreno, J. I., Gascoin, S., Herrero, J., Sproles, E. A., Pons, M., Alonso-Gonz lez, E., Hanich, L., Boudhar, A., Musselman, K. N., Molotch, N. P., Sickman, J., and Pomeroy, J. (2017): Different sensitivities of snowpacks to warming in Mediterranean climate mountain areas, Environmental Research Letters, 12, https://doi.org/10.1088/1748-9326/aa70cb,.

[21] Marchane, A., Jarlan, L., Hanich, L., Boudhar, A., Gascoin, S., Tavernier, A., Filali, N., Le Page, M., Hagolle, O., and Berjamy, B. (2015): Assessment of daily MODIS snow cover products to monitor snow cover dynamics over the Moroccan Atlas mountain range, Remote Sensing of Environment, 160, 72–86, https://doi.org/10.1016/j.rse.2015.01.002.

[22] Marchane, A., Tramblay, Y., Hanich, L., Ruelland, D., and Jarlan, L. (2017): Climate change impacts on surface water resources in the Rheraya catchment (High Atlas, Morocco), Hydrological Sciences Journal, 62, 979–995, https://doi.org/10.1080/02626667.2017.1283042, https: //doi.org/10.1080/02626667.2017.1283042.

[23] Michelangeli, P. A., Vrac, M., and Loukos, H. (2009): Probabilistic downscaling approaches: Application to wind cumulative distribution functions, Geophysical Research Letters, 36, L11 708, https://doi.org/10.1029/2009GL038401, http://doi.wiley.com/10.1029/2009GL038401.

[24] Milewski, A., Elkadiri, R., and Durham, M. (2015): Assessment and comparison of TMPA satellite precipitation products in varying climatic and topographic regimes in Morocco, Remote Sensing, 7, 5697–5717, https://doi.org/10.3390/rs70505697.

[25] Nash, J. E. and Sutcliffe, J. V. (1970): River flow forecasting through conceptual models part I - A discussion of principles, Journal of Hydrology, 10, 282–290, https://doi.org/10.1016/0022-1694(70)90255-6.

[26] Ouatiki, H., Boudhar, A., Tramblay, Y., Jarlan, L., Benabdelouhab, T., Hanich, L., El Meslouhi, M. R., and Chehbouni, A. (2017): Evaluation of TRMM 3B42 V7 rainfall product over the Oum Er Rbia watershed in Morocco, Climate, 5, https://doi.org/10.3390/cli5010001.

[27] Pal, J. S., Giorgi, F., Bi, X., Elguindi, N., Solmon, F., Gao, X., Rauscher, S. A., Francisco, R., Zakey, A., Winter, J., Ashfaq, M., Syed, F. S., Bell, J. L., Differbaugh, N. S., Karmacharya, J., Konari, A., Martinez, D., Da Rocha, R. P., Sloan, L. C., and Steiner, L. (2007): Regional climate modeling for the developing world: The ICTP RegCM3 and RegCNET, Bulletin of the American Meteorological Society, 88, 1395–1409, https://doi.org/10.1175/BAMS-88-9-1395, https://journals.ametsoc.org/bams/article/88/9/1395/59093/ Regional-Climate-Modeling-for-the-Developing-World.

[28] Schulz, O. and de 410 Jong, C. (2004): Snowmelt and sublimation: Field experiments and modelling in the High AtlasMountains ofMorocco, Hydrology and Earth System Sciences, 8, 1076–1089, https://doi.org/10.5194/hess-8-1076-2004.

[29] Steinschneider, S., Yang, Y. C. E., and Brown, C. (2013): Panel regression techniques for identifying impacts of anthropogenic landscape change on hydrologic response, Water

Resources Research, 49, 7874–7886, https://doi.org/10.1002/2013WR013818, http://doi.wiley.com/10.1002/2013WR013818.
[30] Taylor, K. E., Stouffer, R. J., and Meehl, G. A. (2012): An overview of CMIP5 and the experiment design, Bulletin of the American Meteorological Society, 93, 485–498, https://doi.org/10.1175/BAMS-D-11-00094.1.
[31] Tuel, A. and Eltahir, E. A. (2018): Seasonal Precipitation Forecast Over Morocco, Water Resources Research, 54, 9118–9130, https://doi.org/10.1029/2018WR022984.
[32] Tuel, A. and Eltahir, E. A. B. (2020): Why Is the Mediterranean a Climate Change Hot Spot?, Journal of Climate, 33, 5829–5843, https://doi.org/10.1175/JCLI-D-19-0910.1, http://journals.ametsoc.org/doi/10.1175/JCLI-D-19-0910.1https://journals.ametsoc.org/jcli/article/33/14/5829/347612/Why-Is-the-Mediterranean-a-Climate-Change-Hot-Spot.
[33] Tuel, A., Chehbouni, A., and A. B. Eltahir, E. (2020a): Dynamics of seasonal snowpack over the High Atlas, Journal of Hydrology, p. 125657, https://doi.org/10.1016/j.jhydrol.2020.125657, https://linkinghub.elsevier.com/retrieve/pii/S0022169420311185.
[34] Tuel, A., Kang, S., and Eltahir, E. A. B. (2020b): Understanding climate change over the southwestern Mediterranean using high-resolution simulations, Climate Dynamics, https://doi.org/10.1007/s00382-020-05516-8, http://link.springer.com/10.1007/s00382-020-05516-8.
[35] Wan, Z., Hook, S., Hulley, G. (2015): MOD11A1 MODIS/Terra Land Surface Temperature/Emissivity Daily L3 Global 1km SIN Grid V006. NASA EOSDIS Land Processes DAAC, https://doi.org/10.5067/MODIS/MOD11A1.006, https://doi.org/10.5067/MODIS/MOD11A1.006.
[36] Wang, W., Zou, S., Shao, Q., Xing, W., Chen, X., Jiao, X., Luo, Y., Yong, B., and Yu, Z. (2016): The analytical derivation of multiple elasticities of runoff to climate change and catchment characteristics alteration, Journal of Hydrology, 541, 1042–1056, https://doi.org/10.1016/j.jhydrol.2016.08.014, http://www.sciencedirect.com/science/article/pii/S0022169416305042.

Chapter 6

[1] Abatzoglou JT, Drobowski SZ, Parks SA, Hegewisch K (2018) TerraClimate, a high-resolution global dataset of monthly climate and climatic water balance from 1958–2015. Scientific Data 5:170191.
[2] Allan JA, Allan T (2002) The Middle East water question: Hydropolitics and the global economy. I.B. Tauris, London, New York, 382 pp.
[3] Bentsen M, Bethke I, Debernard JB et al (2013) The Norwegian Earth System Model, NorESM1-M – Part 1: Description and basic evaluation of the physical climate. Geoscientific Model Development 6:687-720, https://doi.org/10.5194/gmd-6-687-2013
[4] Bolle HJ (2003) Climate, climate variability, and impacts in the Mediterranean area: An overview. Mediterranean Climate: Variability and Trends, Springer, 1–81.
[5] Brisson E, Van Weverberg K, Demuzere M et al (2016) How well can a

convection-permitting climate model reproduce decadal statistics of precipitation, temperature and cloud characteristics? Clim Dyn 47:3043–3061, https://doi.org/10.1007/s00382-016-3012-z

[6] Bucchignani E, Mercogliano P, Panitz HJ, Montesarchio M (2018) Climate change projections for the Middle East–North Africa domain with COSMO-CLM at different spatial resolutions. Advances in Climate Change Research 9:66–80.

[7] Byrne MP, O'Gorman PA (2015) The Response of Precipitation Minus Evapotranspiration to Climate Warming: Why the "Wet-Get-Wetter, Dry-Get-Drier" Scaling Does Not Hold over Land. J. Climate 28:8078–8092. https://doi.org/10.1175/JCLI-D-15-0369.1.

[8] Cardoso RM, Soares PMM, Miranda PMA, Belo-Pereira M (2013) WRF high resolution simulation of Iberian mean and extreme precipitation climate. Int J Climatol 33:2591–2608, https://doi.org/10.1002/joc.3616

[9] Chan SC, Kendon EJ, Fowler HJ, Blenkinsop S, Ferro CAT, Stephenson DB (2013) Does increasing the spatial resolution of a regional climate model improve the simulated daily precipitation? Clim Dyn 41:1475–1495, https://doi.org/10.1007/s00382-012-1568-9

[10] Chenoweth J, Hadjinicolaou P, Bruggeman A et al (2011) Impact of climate change on the water resources of the eastern Mediterranean and Middle East region: Modeled 21st century changes and implications. Water Resour Res 47:W06506, https://doi.org/10.1029/2010WR010269

[11] Cook BI, Anchukaitis KJ, Touchan R, Meko DM, Cook ER (2016) Spatiotemporal drought variability in the Mediterranean over the last 900 years. J Geophys Res Atmos 121:2060–2074.

[12] FAO (2020) AQUASTAT Core Database. Food and Agriculture Organization of the United Nations. Database accessed on January 21st, 2020.

[13] Fosser G, Khodayar S, Berg P (2015) Benefit of convection permitting climate model simulations in the representation of convective precipitation. Clim Dyn 44(1–2):1–16

[14] Giorgi F, Lionello P (2008) Climate change projections for the Mediterranean region. Global Planet Change 63(2-3):90–104.

[15] Giorgetta MA, Jungclaus J, Reick CH et al (2013) Climate and carbon cycle changes from 1850 to 2100 in MPI-ESM simulations for the Coupled Model Intercomparison Project phase 5. J. Adv Model Earth Syst 5: 572-597.

[16] Gleick PH (1994) Water, War and Peace in the Middle East. Environment: Science and Policy for Sustainable Development, 36(3):6–42. https://doi.org/10.1080/00139157.1994.9929154

[17] Haylock MR, Hofstra N et al (2008) A European daily high-resolution gridded data set of surface temperature and precipitation for 1950–2006. J Geophys Res 113:D20119.

[18] Hochman A, Mercogliano P, Alpert P, Saaroni H, Bucchignani E (2018) High-resolution projection of climate change and extremity over Israel using COSMO-CLM. Int J Climatol 38:5095–5106. https://doi.org/10.1002/joc.5714

[19] Hoerling M, Eischeid J, Perlwitz J et al (2012) On the increased frequency of Mediterranean drought. J Climate 25:2146–2161.

[20] Im ES, Gianotti RL, Eltahir EAB (2014) Improving the simulation of the West African monsoon using the MIT Regional Climate Model. J Clim 27:2209–2229.

[21] Kelley CP, Mohtadi S, Cane MA, Seager R, Kushnir Y (2015) Climate change in the Fertile Crescent and implications of the recent Syrian drought. Proc Natl Acad. Sci USA 112:3241–3246.

[22] Krichak SO, Breitgand JS, Samuels R et al. A double-resolution transient RCM climate change simulation experiment for near-coastal eastern zone of the Eastern Mediterranean region. Theor Appl Climatol 103, 167–195 (2011). https://doi.org/10.1007/s00704-010-0279-6

[23] Kendon E, Ban N, Roberts N, Fowler H, Roberts M, Chan S, Evans J, Fosser G, Wilkinson J (2017) Do convection-permitting regional climate models improve projections of future precipitation change? Bull Amer Meteor Soc 98(1): 79–93, https://doi.org/10.1175/BAMS-D-15-0004.1

[24] Klein Tank AMG, Wijngaard JB, Können GP et al (2002) Daily dataset of 20th-century surface air temperature and precipitation series for the European Climate Assessment. Int J Climatol 22:1441–1453.

[25] Liu C, Ikeda K, Rasmussen R et al (2017) Continental-scale convection-permitting modeling of the current and future climate of North America. Clim Dyn 49:71–95, https://doi.org/10.1007/s00382-016-3327-9

[26] Meehl GA et al (2012) Climate System Response to External Forcings and Climate Change Projections in CCSM4. J Clim 25:3661-3683.

[27] Önol B, Bozkurt D, Turuncoglu UU, Sen OL, Dalfes HN (2014) Evaluation of the twenty-first century RCM simulations driven by multiple GCMs over the Eastern Mediterranean-Black Sea region. Clim Dyn 42:1949–1965.

[28] Ozturk T, Turp MT, Türkes M, Kurnaz ML (2018) Future projections of temperature and precipitation climatology for CORDEX-MENA domain using RegCM4.4. Atmospheric Research 206:87–107.

[29] Pal JS et al (2007) Regional Climate Modeling for the Developing World: The ICTP RegCM3 and RegCNET. Bull Am Meteorol Soc 88:1395–1409.

[30] Pal JS, Eltahir EAB (2016) Future temperature in southwest Asia projected to exceed a threshold for human adaptability. Nature Clim Change 6:128–129

[31] Riahi K et al (2011) RCP8.5 - A scenario of comparatively high greenhouse gas emissions. Climatic Change 109:33-57. https://doi.org/10.1007/s10584-011-0149-y

[32] Ruti P et al (2016) MED-CORDEX initiative for Mediterranean climate studies. Bull Am Meteor Soc 97:1187–1208. https://doi.org/10.1175/BAMS-D-14-

00176.1
[33] Saeed S, Brisson E, Demuzere M et al (2017) Multidecadal convection permitting climate simulations over Belgium: sensitivity of future precipitation extremes. Atmos Sci Lett 18:29–36. https://doi.org/10.1002/asl.720
[34] Scheffran J, Battaglini A (2011) Climate and conflicts: the security risks of global warming. Reg Environ Change 11:27–39.
[35] Tuel A, Eltahir EAB (2020) Why is the Mediterranean a Climate Change Hotspot? J. Climate, 33, 5829–5843, https://doi.org/10.1175/JCLI-D-19-0910.1.
[36] Thomson A et al (2011) RCP4.5: a pathway for stabilization of radiative forcing by 2100. Climatic Change 109:77-94. https://doi.org/10.1007/s10584-011-0151-4
[37] Winter JM, Pal JS, Eltahir EAB (2009) Coupling of Integrated Biosphere Simulator to Regional Climate Model Version 3. J Clim 22:2743–2757.
[38] Zappa G, Hawcroft MK, Shaffrey L, Black E, Brayshaw DJ (2015) Extratropical cyclones and the projected decline of winter Mediterranean precipitation in the CMIP5 models. Clim Dyn 45:1727–1738.
[39] Zittis G, Hadjinicolaou P, Lelieveld J (2014) Comparison of WRF Model physics parameterizations over the MENA-CORDEX domain. Amer J Clim Change 3:490-511. https://doi.org/10.4236/ajcc.2014.35042
[40] Zittis G, Bruggeman A, Camera C, Hadjinicolaou P, Lelieveld J (2017) The added value of convection permitting simulations of extreme precipitation events over the eastern mediterranean. Atmos Res 191:20–33. https://doi.org/10.1016/j.atmosres.2017.03.002

Chapter 7

[1] Adler, R. F., and Coauthors, (2003): The version-2 Global Precipitation Climatology Project (GPCP) monthly precipitation analysis (1979–present). J. Hydrometeor., 4, 1147–1167, https://doi.org/10.1175/1525-7541(2003)004<1147:TVGPCP>2.0.CO;2.
[2] Bador, M., L. Terray, and J. Boé, (2016): Emergence of human influence on summer record–breaking temperatures over Europe. Geophys. Res. Lett., 43, 404–412, https://doi.org/10.1002/2015GL066560.
[3] Belleflamme, A., X. Fettweis, and M. Erpicum, (2015): Do global warming–induced circulation pattern changes affect temperature and precipitation over Europe during summer? Int. J. Climatol., 35, 1484–1499, https://doi.org/10.1002/joc.4070.
[4] Bladé, I., D. Fortuny, G. J. van Oldenborgh, and B. Liebmann, (2012): The summer North Atlantic Oscillation in CMIP3 models and related uncertainties in projected summer drying in Europe. J. Geophys. Res. Atmos., 117, D16104, https://doi.org/10.1029/2012JD017816.
[5] Boberg, F., and J. H. Christensen, (2012): Overestimation of Mediterranean summer temperature projections due to model deficiencies. Nat. Climate Change, 2, 433–436, https://doi.org/10.1038/nclimate1454.

[6] Boé, J., and L. Terray, (2014): Land–sea contrast, soil-atmosphere and cloud-temperature interactions: Interplays and roles in future summer European climate change. Climate Dyn., 42, 683–699, https://doi.org/10.1007/s00382-013-1868-8.

[7] Boé, J., L. Terray, C. Cassou, and J. Najac, (2009): Uncertainties in European summer precipitation changes: Role of large scale circulation. Climate Dyn., 33, 265–276, https://doi.org/10.1007/s00382-008-0474-7.

[8] Boé, J., S. Somot, L. Corre, and P. Nabat, (2020a): Large discrepancies in summer climate change over Europe as projected by global and regional climate models: Causes and consequences. Climate Dyn., 54, 2981–3002, https://doi.org/10.1007/s00382-020-05153-1.

[9] Boé, J., and Coauthors, (2020b): Past long-term summer warming over western Europe in new generation climate models: Role of large-scale atmospheric circulation. Environ. Res. Lett., 15, 084038, https://doi.org/10.1088/1748-9326/ab8a89.

[10] Brogli, R., S. L. Sørland, N. Kröner, and C. Schär, (2019): Causes of future Mediterranean precipitation decline depend on the season. Environ. Res. Lett., 14, 114017, https://doi.org/10.1088/1748-9326/ab4438.

[11] Byrne, M. P., and P. A. O'Gorman, (2018): Trends in continental temperature and humidity directly linked to ocean warming. Proc. Natl. Acad. Sci. USA, 115, 4863–4868, https://doi.org/10.1073/pnas.1722312115.

[12] Cornes, R. C., G. van der Schrier, E. J. M. van den Besselaar, and P. D. Jones, (2018): An ensemble version of the E-OBS temperature and precipitation data sets. J. Geophys. Res. Atmos., 123, 9391–9409, https://doi.org/10.1029/2017JD028200.

[13] Dong, B., R. T. Sutton, T. Woollings, and K. Hodges, (2013): Variability of the North Atlantic summer storm track: Mechanisms and impacts on European climate. Environ. Res. Lett., 8, 034037, https://doi.org/10.1088/1748-9326/8/3/034037.

[14] Eltahir, E. A. B., (1998): A soil moisture–rainfall feedback mechanism: 1. Theory and observations. Water Resour. Res., 34, 765–776, https://doi.org/10.1029/97WR03499.

[15] Fischer, E. M., S. I. Seneviratne, P. L. Vidale, D. Lüthi, and C. Schär, (2007): Soil moisture–atmosphere interactions during the 2003 European summer heat wave. J. Climate, 20, 5081–5099, https://doi.org/10.1175/JCLI4288.1.

[16] Gervais, M., J. Shaman, and Y. Kushnir, (2019): Impacts of the North Atlantic warming hole in future climate projections: Mean atmospheric circulation and the North Atlantic jet. J. Climate, 32, 2673–2689, https://doi.org/10.1175/JCLI-D-18-0647.1.

[17] Grillakis, M. G., (2019): Increase in severe and extreme soil moisture droughts for Europe under climate change. Sci. Total Environ., 660, 1245–1255, https://doi.org/10.1016/j.scitotenv.2019.01.001.

[18] Gruber, A., T. Scanlon, R. van der Schalie, W. Wagner, and W. Dorigo, (2019): Evolution of the ESA CCI soil moisture climate data records and their underlying merging methodology. Earth Syst. Sci. Data, 11, 717–739, https://doi.org/10.5194/essd-11-717-2019.

[19] Haarsma, R. J., F. Selten, B. van den Hurk, W. Hazeleger, and X. Wang, (2009): Drier Mediterranean soils due to greenhouse warming bring easterly winds over summertime central Europe. Geophys. Res. Lett., 36, L04705, https://doi.org/10.1029/2008GL036617.

[20] Haarsma, R. J., F. Selten, and S. S. Drijfhout, (2015): Decelerating Atlantic meridional overturning circulation main cause of future west European summer atmospheric

circulation changes. Environ. Res. Lett., 10, 094007, https://doi.org/10.1088/1748-9326/10/9/094007.

[21] Harris, I., T. J. Osborn, P. Jones, and D. Lister, (2020): Version 4 of the CRU TS monthly high-resolution gridded multivariate climate dataset. Sci. Data, 7, 109, https://doi.org/10.1038/s41597-020-0453-3.

[22] Hersbach, H., and Coauthors, (2020): The ERA5 global reanalysis. Quart. J. Roy. Meteor. Soc., 146, 1999–2049, https://doi.org/10.1002/qj.3803.

[23] Hohenegger, C., P. Brockhaus, C. S. Bretherton, and C. Schär, (2009): The soil moisture–precipitation feedback in simulations with explicit and parameterized convection. J. Climate, 22, 5003–5020, https://doi.org/10.1175/2009JCLI2604.1.

[24] Hu, S., and A. V. Fedorov, (2020): Indian Ocean warming as a driver of the North Atlantic warming hole. Nat. Commun., 11, 4785, https://doi.org/10.1038/s41467-020-18522-5.

[25] Im, E. S., R. L. Gianotti, and E. A. B. Eltahir, (2014): Improving the simulation of the West African monsoon using the MIT regional climate model. J. Climate, 27, 2209–2229, https://doi.org/10.1175/JCLI-D-13-00188.1.

[26] Kew, S., S. Y. Philip, G. Jan van Oldenborgh, G. van der Schrier, F. E. L. Otto, and R. Vautard, (2019): The exceptional summer heat wave in southern Europe 2017. Bull. Amer. Meteor. Soc., 100 (1), S49–S53, https://doi.org/10.1175/BAMS-D-18-0109.1.

[27] Knist, S., and Coauthors, (2017): Land–atmosphere coupling in EURO-CORDEX evaluation experiments. J. Geophys. Res. Atmos., 122, 79–103, https://doi.org/10.1002/2016JD025476.

[28] Lionello, P., and L. Scarascia, (2018): The relation between climate change in the Mediterranean region and global warming. Reg. Environ. Change, 18, 1481–1493, https://doi.org/10.1007/s10113-018-1290-1.

[29] Mueller, B., and S. I. Seneviratne, (2012): Hot days induced by precipitation deficits at the global scale. Proc. Natl. Acad. Sci. USA, 109, 12 398–12 403, https://doi.org/10.1073/pnas.1204330109.

[30] Orth, R., and S. I. Seneviratne, (2017): Variability of soil moisture and sea surface temperatures similarly important for warm-season land climate in the Community Earth System Model. J. Climate, 30, 2141–2162, https://doi.org/10.1175/JCLI-D-15-0567.1.

[31] Orth, R., J. Zscheischler, and S. I. Seneviratne, (2016): Record dry summer in 2015 challenges precipitation projections in Central Europe. Sci. Rep., 6, 28334, https://doi.org/10.1038/srep28334.

[32] Pal, J. S., and Coauthors, (2007): Regional climate modeling for the developing world: The ICTP RegCM3 and RegCNET. Bull. Amer. Meteor. Soc., 88, 1395–1410, https://doi.org/10.1175/BAMS-88-9-1395.

[33] Quesada, B., R. Vautard, P. Yiou, M. Hirschi, and S. I. Seneviratne, (2012): Asymmetric European summer heat predictability from wet and dry southern winters and springs. Nat. Climate Change, 2, 736–741, https://doi.org/10.1038/nclimate1536.

[34] Rodwell, M. J., and B. J. Hoskins, (2001): Subtropical anticyclones and summer monsoons. J. Climate, 14, 3192–3211, https://doi.org/10.1175/1520-0442(2001)014<3192:SAASM>2.0.CO;2.

[35] Rowell, D. P., and R. G. Jones, (2006): Causes and uncertainty of future summer drying over Europe. Climate Dyn., 27, 281–299, https://doi.org/10.1007/s00382-006-0125-9.

[36] Samaniego, L., and Coauthors, (2018): Anthropogenic warming exacerbates European soil moisture droughts. Nat. Climate Change, 8, 421–426, https://doi.org/10.1038/s41558-018-0138-5.
[37] Sandler, D., and N. Harnik, (2020): Future wintertime meridional wind trends through the lens of subseasonal teleconnections. Wea. Climate Dyn., 1, 427–443, https://doi.org/10.5194/wcd-1-427-2020.
[38] Schär, C., P. L. Vidale, D. Lüthi, C. Frei, C. Häberli, M. A. Liniger, and C. Appenzeller, (2004): The role of increasing temperature variability in European summer heatwaves. Nature, 427, 332–336, https://doi.org/10.1038/nature02300.
[39] Seager, R., H. Liu, N. Henderson, I. R. Simpson, C. Kelley, T. Shaw, Y. Kushnir, and M. Ting, (2014): Causes of increasing aridification of the Mediterranean region in response to rising greenhouse gases. J. Climate, 27, 4655–4676, https://doi.org/10.1175/JCLI-D-13-00446.1.
[40] Seneviratne, S. I., D. Lüthi, M. Litschi, and C. Schär, (2006): Land–atmosphere coupling and climate change in Europe. Nature, 443, 205–209, https://doi.org/10.1038/nature05095.
[41] Seneviratne, S. I., T. Corti, E. L. Davin, M. Hirschi, E. B. Jaeger, I. Lehner, B. Orlowsky, and A. J. Teuling, (2010): Investigating soil moisture–climate interactions in a changing climate: A review. Earth-Sci. Rev., 99, 125–161, https://doi.org/10.1016/j.earscirev.2010.02.004.
[42] Simpson, I. R., R. Seager, M. Ting, and T. A. Shaw, (2016): Causes of change in Northern Hemisphere winter meridional winds and regional hydroclimate. Nat. Climate Change, 6, 65–70, https://doi.org/10.1038/nclimate2783.
[43] Stott, P. A., D. A. Stone, and M. R. Allen, (2004): Human contribution to the European heatwave of 2003. Nature, 432, 610–614, https://doi.org/10.1038/nature03089.
[44] Sutton, R. T., B. Dong, and J. M. Gregory, (2007): Land/sea warming ratio in response to climate change: IPCC AR4 model results and comparison with observations. Geophys. Res. Lett., 34, L02701, https://doi.org/10.1029/2006GL028164.
[45] Taylor, K. E., R. J. Stouffer, and G. A. Meehl, (2012): An overview of CMIP5 and the experiment design. Bull. Amer. Meteor. Soc., 93, 485–498, https://doi.org/10.1175/BAMS-D-11-00094.1.
[46] Tuel, A., and E. A. B. Eltahir, (2020): Why is the Mediterranean a climate change hot spot? J. Climate, 33, 5829–5843, https://doi.org/10.1175/JCLI-D-19-0910.1.
[47] Tuel, A., S. Kang, and E. A. B. Eltahir, (2021a): Understanding climate change over the southwestern Mediterranean using high-resolution simulations. Climate Dyn., 56, 985–1001, https://doi.org/10.1007/s00382-020-05516-8.
[48] Tuel, A., P. A. O'Gorman, and E. A. B. Eltahir, (2021b): Elements of the dynamical response to climate change over the Mediterranean. J. Climate, 34, 1135–1146, https://doi.org/10.1175/JCLI-D-20-0429.1.
[49] Vautard, R., and Coauthors, (2007): Summertime European heat and drought waves induced by wintertime Mediterranean rainfall deficit. Geophys. Res. Lett., 34, L07711, https://doi.org/10.1029/2006GL028001.
[50] Vautard, R., and Coauthors, (2020): Human contribution to the record-breaking June and July 2019 heatwaves in Western Europe. Environ. Res. Lett., 15, 094077, https://doi.org/10.1088/1748-9326/aba3d4.

[51] Wang, G., A. J. Dolman, and A. Alessandri, (2011): A summer climate regime over Europe modulated by the North Atlantic Oscillation. Hydrol. Earth Syst. Sci., 15, 57–64, https://doi.org/10.5194/hess-15-57-2011.

[52] Wilks, D. S., (2016): "The stippling shows statistically significant grid points": How research results are routinely overstated and overinterpreted, and what to do about it. Bull. Amer. Meteor. Soc., 97, 2263–2273, https://doi.org/10.1175/BAMS-D-15-00267.1.

[53] Wills, R. C., R. H. White, and X. J. Levine, (2019): Northern Hemisphere stationary waves in a changing climate. Curr. Climate Change Rep., 5, 372–389, https://doi.org/10.1007/s40641-019-00147-6.

[54] Zappa, G., (2019): Regional climate impacts of future changes in the mid-latitude atmospheric circulation: A storyline view. Curr. Climate Change Rep., 5, 358–371, https://doi.org/10.1007/s40641-019-00146-7.

Made in the USA
Monee, IL
05 November 2026

67858074R00127